What a Waste 2.0

Urban Development Series

The Urban Development Series discusses the challenge of urbanization and what it will mean for developing countries in the decades ahead. The series aims to delve substantively into a range of core issues related to urban development that policy makers and practitioners must address.

Cities and Climate Change: Responding to an Urgent Agenda

Climate Change, Disaster Risk, and the Urban Poor: Cities Building Resilience for a Changing World

East Asia and Pacific Cities: Expanding Opportunities for the Urban Poor

East Asia's Changing Urban Landscape: Measuring a Decade of Spatial Growth

The Economics of Uniqueness: Investing in Historic City Cores and Cultural Heritage Assets for Sustainable Development

Financing Transit-Oriented Development with Land Values: Adapting Land Value Capture in Developing Countries

Regenerating Urban Land: A Practitioner's Guide to Leveraging Private Investment

Transforming Cities with Transit: Transit and Land-Use Integration for Sustainable Urban Development

Urban Risk Assessments: Understanding Disaster and Climate Risk in Cities

What a Waste 2.0: A Global Snapshot of Solid Waste Management to 2050

All books in the Urban Development Series are available free at
https://openknowledge.worldbank.org/handle/10986/2174.

What a Waste 2.0

A Global Snapshot of Solid Waste Management to 2050

Silpa Kaza, Lisa Yao, Perinaz Bhada-Tata, and
Frank Van Woerden

With Kremena Ionkova, John Morton,
Renan Alberto Poveda, Maria Sarraf, Fuad Malkawi,
A.S. Harinath, Farouk Banna, Gyongshim An,
Haruka Imoto, and Daniel Levine

Contents

Foreword   xi

Acknowledgments   xiii

Abbreviations   xvii

1   Introduction   1
    A Note on Data   9
    Notes   13
    References   13

2   At a Glance: A Global Picture of
    Solid Waste Management   17
    Key Insights   17
    Waste Generation   18
    Projected Waste Generation   24
    Waste Composition   29
    Waste Collection   32
    Waste Disposal   34
    Special Wastes   36
    Notes   37
    References   37

3   Regional Snapshots   39
    East Asia and Pacific   39
    Europe and Central Asia   46
    Latin America and the Caribbean   53

Middle East and North Africa   59
North America   66
South Asia   69
Sub-Saharan Africa   76
References   83
Additional Resources   84

4   Waste Administration and Operations   87
Key Insights   87
Solid Waste Regulations   89
Solid Waste Planning   91
Institutions and Coordination   93
Waste Management Operations   94
References   99

**5   Financing and Cost Recovery for Waste
      Management Systems   101**
Key Insights   101
Waste Management Budgets   102
Waste Management Costs   103
Waste Management Financing   105
References   112

6   Waste and Society   115
Key Insights   115
Environment and Climate Change   116
Technology Trends   121
Citizen Engagement   126
Social Impacts of Waste Management and the Informal Sector   129
Notes   133
References   133
Additional Resources   139

7   Case Studies   141
1. A Path to Zero Waste in San Francisco, United States   141
2. Achieving Financial Sustainability in Argentina and Colombia   143
3. Automated Waste Collection in Israel   147
4. Cooperation between National and Local Governments for
       Municipal Waste Management in Japan   148
5. Central Reforms to Stabilize the Waste Sector and Engage the
       Private Sector in Senegal   151
6. Decentralized Organic Waste Management by Households in
       Burkina Faso   152
7. Eco-Lef: A Successful Plastic Recycling System in Tunisia   153
8. Extended Producer Responsibility Schemes in Europe   155
9. Financially Resilient Deposit Refund System: The Case of the
       Bottle Recycling Program in Palau   158

10. Improving Waste Collection by Partnering with the Informal
     Sector in Pune, India   161
11. Improving Waste Management through Citizen Communication
     in Toronto, Canada   163
12. Managing Disaster Waste   165
13. Minimizing Food Loss and Waste in Mexico   167
14. Sustainable Source Separation in Panaji, India   170
15. Musical Garbage Trucks in Taiwan, China   173
16. The Global Tragedy of Marine Litter   174
17. Using Information Management to Reduce Waste in Korea   176
Notes   177
References   177
Additional Resources   180

Appendix A   Waste Generation (tonnes per year) and
              Projections by Country or Economy   185

Appendix B   Waste Treatment and Disposal
              by Country or Economy   231

Boxes

1.1   Data for the Sub-Saharan Africa Region   10
2.1   Waste Generation Projection Methodology   25
2.2   Global Food Loss and Waste   30
3.1   Morocco: Investing in Environmental Sustainability Pays Off   60
3.2   Swachh Bharat Mission (Clean India Mission)   75
5.1   Results-Based Financing in Waste Management   110
5.2   Carbon Finance   111
6.1   Plastic Waste Management   117
6.2   Examples of Information That Can Be Aggregated Using a
      Waste Management Data System   122
6.3   I Got Garbage   124
6.4   Mr. Trash Wheel   127
6.5   Waste Picker Cooperative Model: Recuperar   130
6.6   Formalization of Waste Pickers in Brazil   131
6.7   Challenges for Waste Pickers   131
6.8   Socially Responsible Plastics Recycling in Mexico   132

Figures

2.1   Waste Generation by Region   19
2.2   Waste Generation by Income Level   21
2.3   Waste Generation and Gross Domestic Product   22
2.4   Waste Generation and Urbanization Rate   23
2.5   Projected Global Waste Generation   25
B2.1.1 Waste Generation: Actual and Model Prediction   26
2.6   Projected Waste Generation by Income Group   27
2.7   Projected Waste Generation by Region   28

2.8	Global Waste Composition	29
2.9	Waste Composition by Income Level	30
2.10	Waste Collection Rates	32
2.11	Urban and Rural Collection Rates by Income Level	33
2.12	Global Waste Treatment and Disposal	34
2.13	Disposal Methods by Income	35
2.14	Global Average Special Waste Generation	36
3.1	Waste Generation Rates: East Asia and Pacific Region	40
3.2	Waste Composition in East Asia and Pacific	41
3.3	Waste Collection Coverage in East Asia and Pacific	42
3.4	Waste Collection Rates for Select Cities in East Asia and Pacific	44
3.5	Number of Cities in East Asia and Pacific Source Separating Recoverable Waste Streams	45
3.6	Waste Disposal and Treatment in East Asia and Pacific	45
3.7	Waste Generation Rates: Europe and Central Asia	47
3.8	Waste Composition in Europe and Central Asia	47
3.9	Waste Collection Coverage in Europe and Central Asia	48
3.10	Waste Collection Rates for Select Cities in Europe and Central Asia	49
3.11	Number of Cities in Europe and Central Asia Source Separating Recoverable Waste Streams	50
3.12	Waste Disposal and Treatment in Europe and Central Asia	51
3.13	Waste Generation Rates: Latin America and the Caribbean Region	54
3.14	Waste Composition in Latin America and the Caribbean	54
3.15	Waste Collection Coverage in Latin America and the Caribbean	55
3.16	Waste Collection Rates for Select Cities in Latin America and the Caribbean	56
3.17	Waste Disposal and Treatment in Latin America and the Caribbean	58
3.18	Waste Generation Rates: Middle East and North Africa Region	60
3.19	Waste Composition in the Middle East and North Africa	61
3.20	Waste Collection Coverage in the Middle East and North Africa	62
3.21	Waste Collection Rates for Select Cities in the Middle East and North Africa	63
3.22	Waste Disposal and Treatment in the Middle East and North Africa	65
3.23	Waste Generation Rates: North American Region	67
3.24	Waste Composition in North America	67
3.25	Waste Disposal and Treatment in North America	68
3.26	Waste Generation Rates: South Asia Region	70
3.27	Waste Composition in South Asia	71
3.28	Waste Collection Coverage in South Asia	71
3.29	Waste Collection Rates for Select Cities in South Asia	72

3.30   Waste Collection Methods in South Asia   73
3.31   Waste Disposal and Treatment in South Asia   75
3.32   Waste Generation Rates: Sub-Saharan Africa Region   78
3.33   Waste Composition in Sub-Saharan Africa   78
3.34   Waste Collection Coverage in Sub-Saharan Africa   79
3.35   Waste Collection Rates for Select Cities in Sub-Saharan Africa   80
3.36   Waste Disposal and Treatment in Sub-Saharan Africa   82
4.1   Waste Management Administration, Operation, and
        Financing Models   97
5.1   Waste Management Fee Type and Billing Method   108
B6.3.1 Features of I Got Garbage Application   124
7.2.1   Cost Recovery by Generator in Argentina   144
7.2.2   Urban Solid Waste Management Costs (US$) by Stage   144
7.2.3   Urban Solid Waste Management by Spending Category
        in Argentina   145
7.11.1 Screenshot of Waste Wizard on the City of Toronto Website   163

Maps

1.1   Definition of Income Levels   7
1.2   Definition of Regions   8
2.1   Waste Generation Per Capita   19

Photos

1.1   Plastic Waste at the Thilafushi Waste Disposal Site, Maldives   3
1.2   A Recycler Transports Waste Using a Modified Motorcycle,
        Bangkok, Thailand   4
3.1   Landfill in China   43
3.2   A Waste and Street Cleaning Worker in Hoi An, Vietnam   43
3.3   Recycling Plant in Bosnia and Herzegovina   51
3.4   Plastic Bottle Collection in Jamaica   57
3.5   One Form of Waste Collection in West Bank   62
3.6   Dumpsite in Sri Lanka   74
3.7   Waste Collectors in Uganda   81
4.1   Plastic Bag Ban in Kenya   90
6.1   Solar-Powered Waste Compaction Bins in the Czech Republic   123
6.2   Informal Recyclers in the Middle East and North Africa Region   129
7.1   Japanese Bins   150
7.2   Japanese Recycling Facility   150
7.3   Eco-Lef Workers Collecting and Weighing Packaging Waste
        at the Montplaisir Collection Center in Tunis, Tunisia   154
7.4   An Automated Bottle Deposit Machine   157
7.5   Compacting Beverage Containers inside the Plant in Palau   159
7.6   Recovery Efforts after Meethotamulla Dumpsite Collapse
        from Heavy Rains in Colombo, Sri Lanka   166

7.7   Organic Waste Bin in Mexico City, Mexico   168
7.8a and b Sorting Center at Residential Colony in Panaji, India   171
7.8c   Decentralized Composting Units in Panaji, India   171
7.9   Spilled Garbage on the Beach   174

Tables

2.1   Ranges of Average National Waste Generation by Region   22
2.2   Industrial and Electronic Waste Generation Rates   36
3.1   Countries with High Recycling and Composting Rates in
           Europe and Central Asia   52
3.2   Examples of Transfer Station Availability and Transportation
           Distance in the Middle East and North Africa   64
4.1   Existence of National Waste Management Regulation   89
4.2   Existence of Urban Waste Management Regulation   91
4.3   Existence and Implementation of Urban Master Plan   92
4.4   Oversight of Solid Waste Management in Cities   94
4.5   Examples of Waste Management Operations and
           Administrative Models   95
5.1   Solid Waste Management as a Percentage of Municipal Budget   102
5.2   Typical Waste Management Costs by Disposal Type   104
5.3   Capital and Operational Expenditures of Incineration and
           Anaerobic Digestion Systems   105
5.4   Waste Management User Fees by Region   106
5.5   Waste Management User Fees by Income Level   107
7.4.1   Cooperation of National and Local Governments in Japan
           on Municipal Solid Waste Management   148
7.8.1   Number of European Union Member States Implementing
           Extended Producer Responsibility Schemes in 2013   155
7.12.1   Typical Phases of Disaster Waste Management   165

Foreword

As you will see in this report, the world is on a trajectory where waste generation will drastically outpace population growth by more than double by 2050. Although we are seeing improvements and innovations in solid waste management globally, it is a complex issue and one that we need to take urgent action on.

Solid waste management affects everyone; however, those most affected by the negative impacts of poorly managed waste are largely society's most vulnerable—losing their lives and homes from landslides of waste dumps, working in unsafe waste-picking conditions, and suffering profound health repercussions.

Too often, the environment also pays a high price. In 2016, the world generated 242 million tonnes of plastic waste—12 percent of all municipal solid waste. Plastic waste is choking our oceans, yet our consumption of plastics is only increasing. Cities and countries are rapidly developing without adequate systems in place to manage the changing waste composition of citizens.

Meanwhile, an estimated 1.6 billion tonnes of carbon dioxide–equivalent (CO_2-equivalent) greenhouse gas emissions were generated from solid waste management in 2016. This is about 5 percent of global emissions. Without improvements in the sector, solid waste–related emissions are anticipated to increase to 2.6 billion tonnes of CO_2-equivalent by 2050. More than 80 countries committed to reduce emissions through the historic 2017 Paris Agreement—improving waste management is one way of contributing to this effort.

Solid waste management is a critical—yet often overlooked—piece for planning sustainable, healthy, and inclusive cities and communities for all. However, waste management can be the single highest budget item for many local administrations. Municipalities in low-income countries are spending about 20 percent of their budgets on waste management, on average—yet over 90 percent of waste in low-income countries is still openly dumped or burned. As these cities and countries grow rapidly, they desperately need systems to manage their growing waste and mechanisms to pay for the essential services that keep their citizens healthy and their communities clean.

We need cities and countries to plan holistically and manage our precious resources better than we have in the past. This report shows what governments around the world have done to manage their solid waste and highlights the latest trends across income levels and geographies. Building on *What a Waste: A Global Review of Solid Waste Management* from 2012, this report highlights the overwhelming cost of waste management and the need for solutions.

Using the rich findings and data from this report, I urge stakeholders to think ahead and to integrate waste management into their paradigm of economic growth and innovation. It is the responsibility of every citizen, government, business, city, and country to create the healthy, inclusive, and livable shared world that we strive for.

Ede Ijjasz-Vasquez
Senior Director
Social, Urban, Rural and Resilience Global Practice
The World Bank

Acknowledgments

What a Waste 2.0: A Global Snapshot of Solid Waste Management to 2050 was authored by a core team composed of Silpa Kaza, Lisa Yao, Perinaz Bhada-Tata, and Frank Van Woerden. The study was led by Silpa Kaza (Task Team Leader), Frank Van Woerden (Co–Task Team Leader), and Daniel Levine (Co–Task Team Leader). This effort was generously funded by the government of Japan through the World Bank's Tokyo Development Learning Center (TDLC). Daniel Levine and Haruka Imoto represented the TDLC and provided strategic guidance and administrative support from the project's design to finalization. The study was prepared by the World Bank's Social, Urban, Rural and Resilience Global Practice in collaboration with the Environment Global Practice.

The team thanks the following individuals for their valuable guidance: Paul Kriss for advising the team from the beginning, Philip Karp for advising the team on the study and dissemination strategy, Stephane Hallegatte for his guidance on the projections methodology, Mersedeh TariVerdi for her robust modeling of waste generation projections, and Catalina Marulanda for advising the team on technical content.

Each section of the report was written by the following individuals:

- Chapter 1 (Introduction) was written by Silpa Kaza and Frank Van Woerden
- Chapter 2 (At a Glance: A Global Picture of Solid Waste Management) was written by Silpa Kaza and Lisa Yao
- Chapter 3 (Regional Snapshots) was written by Silpa Kaza and Lisa Yao with inputs from regional focal points
- Chapter 4 (Waste Administration and Operations) was written by Lisa Yao with contributions from Frank Van Woerden
- Chapter 5 (Financing and Cost Recovery for Waste Management Systems) was written by Lisa Yao with contributions from Frank Van Woerden

- Chapter 6 (Waste and Society) was written by Silpa Kaza, Lisa Yao, and Perinaz Bhada-Tata with contributions from Frank Van Woerden
- Chapter 7 (Case Studies) was edited by Lisa Yao and Silpa Kaza and written by Perinaz Bhada-Tata, Thierry Martin, Kevin Serrona, Ritu Thakur, Flaviu Pop, Shiko Hayashi, Gustavo Solorzano, Nadya Selene Alencastro Larios, Renan Alberto Poveda, and Anis Ismail

The data collection efforts and case studies were led by World Bank solid waste experts serving as regional focal points along with support from consultants collecting and documenting information on solid waste management data and practices globally. Each regional team was structured as follows:

- *Latin America and the Caribbean*: John Morton and Renan Alberto Poveda served as the regional focal points, with data collection and case study support from Nadya Selene Alencastro Larios, Cauam Cardoso, Bernardo Deregibus, and Gustavo Solorzano.
- *Sub-Saharan Africa*: Gyongshim An and Farouk Mollah Banna served as the regional focal points, with data collection and case study support from Dede Raissa Adomayakpor, Thierry Martin, and Emily Sullivan.
- *East Asia and Pacific*: Frank Van Woerden served as the regional focal point, with data collection and case study support from Kevin Serrona.
- *South Asia*: A.S. Harinath served as the regional focal point, with data collection and case study support from Ritu Thakur.
- *Middle East and North Africa*: Fuad Malkawi and Maria Sarraf served as the regional focal points, with data collection and case study support from Anis Ismail, Omar Ouda, and Ali Abedini.
- *Europe and Central Asia*: Kremena Ionkova served as the regional focal point, with data collection and case study support from Flaviu Pop.
- *Japan*: Haruka Imoto served as the focal point, with data collection and case study support from Shiko Hayashi.
- *High-income countries*: Perinaz Bhada-Tata served as the focal point for remaining high-income countries and provided data collection and case study support. Madhumitha Raj assisted in the collection of data.

Perinaz Bhada-Tata oversaw the data management and validation process, and all data and sources were validated by Perinaz Bhada-Tata, Madhumitha Raj, and Henry Shull. James Michelsen was paramount to finalizing the selection of data metrics. Rubaina Anjum provided content on food loss and waste for the report.

Mersedeh TariVerdi supported the modeling of the projections and developed the regression model under the guidance of Stephane Hallegatte. The methodology for projections benefited greatly from the technical expertise of Paolo Avner and analytical support from Lisa Yao.

Tony Fujs and Meera Desai from the World Bank's Development Economics Group created the database for all waste management data collected and designed the interactive website to host the data.

The study was prepared under the guidance of Ede Ijjasz-Vasquez, senior director of the Social, Urban, Rural and Resilience Global Practice; Sameh Wahba, director of Urban and Territorial Development, Disaster Risk Management, and Resilience; Maitreyi Das, practice manager; and Senait Assefa, practice manager.

Peer reviewers from numerous organizations provided critical expert comments. The team thanks Stephen Hammer, practice manager of Climate Analytics and Advisory Services; Catalina Marulanda, practice manager of the South Asia Urban Unit; Daniel Hoornweg, author of *What a Waste: A Global Review of Solid Waste Management* and professor and research chair at the University of Ontario Institute of Technology; Fabien Mainguy, senior project manager at Suez Environnement; James Law, International Solid Waste Association Board member and project director at SCS Engineers; and Makoto Mihara, director of the Environment Bureau in Osaka.

Tokyo Development Learning Center

The TDLC program is a partnership of Japan and the World Bank. TDLC supports and facilitates strategic World Bank Group and client country collaboration with select Japanese cities, agencies, and partners for joint research, knowledge exchange, capacity building, and other activities that develop opportunities to link Japanese and global expertise with specific project-level engagements in developing countries to maximize development impact.

Abbreviations

ANGed	National Agency for Waste Management
AVAC	automated vacuum collection
BAMX	Mexican Food Banking Network
BOO	build-operate-own
BOT	build-operate-transfer
C&D	construction and demolition
CDM	Clean Development Mechanism
CFB	circulating fluidized bed
CO_2	carbon dioxide
DBFO	design-build-finance-operate
DBO	design-build-operate
DBOT	design-build-operate-transfer
EAP	East Asia and Pacific
ECA	Europe and Central Asia
EPR	extended producer responsibility
EU	European Union
FLW	food loss and waste
GCC	Gulf Cooperation Council
GDP	gross domestic product
GHG	greenhouse gas
GWP	global warming potential
HIC	high-income country
IFC	International Finance Corporation
INR	Indian rupees
JWMA	Japan Waste Management Association
kg	kilogram
LAC	Latin America and the Caribbean
LDPE	low-density polyethylene

LIC	low-income country
LMIC	lower-middle-income country
MENA	Middle East and North Africa
MPIIC	Ministry of Public Infrastructures, Industries and Commerce
MSW	municipal solid waste
MWh	megawatt hour of energy
NA	North America
OECD	Organisation for Economic Co-operation and Development
PET	polyethylene terephthalate
PMC	Pune Municipal Corporation
PPP	public-private partnership
PRO	producer responsibility organization
RBF	results-based financing
RFID	radio-frequency identification chips
SAR	South Asia
SAR	special administrative region
SAyDS	Secretariat of Environment and Sustainable Development
SSA	Sub-Saharan Africa
SWaCH	Solid Waste Collection and Handling or, officially, SWaCH Seva Sahakari Sanstha Maryadit, Pune
UN	United Nations
UNFCCC	United Nations Framework Convention on Climate Change

CHAPTER 1

Introduction

Solid waste management is a universal issue affecting every single person in the world. Individuals and governments make decisions about consumption and waste management that affect the daily health, productivity, and cleanliness of communities. Poorly managed waste is contaminating the world's oceans, clogging drains and causing flooding, transmitting diseases via breeding of vectors, increasing respiratory problems through airborne particles from burning of waste, harming animals that consume waste unknowingly, and affecting economic development such as through diminished tourism. Unmanaged and improperly managed waste from decades of economic growth requires urgent action at all levels of society.

As countries develop from low-income to middle- and high-income levels, their waste management situations also evolve. Growth in prosperity and movement to urban areas are linked to increases in per capita generation of waste. Furthermore, rapid urbanization and population growth create larger population centers, making the collection of all waste and the procuring of land for treatment and disposal more and more difficult.

Urban waste management is expensive. Waste management can be the single highest budget item for many local administrations in low-income countries, where it comprises nearly 20 percent of municipal budgets, on average. In middle-income countries, solid waste management typically accounts for more than 10 percent of municipal budgets, and it accounts for about 4 percent in high-income countries. Budget resources devoted to waste management can be much higher in certain cases.

Costly and complex waste operations must compete for funding with other priorities such as clean water and other utilities, education, and health care. Waste management is often administered by local authorities with limited resources and limited capacity for planning, contract management, and operational monitoring. These factors make sustainable waste management a complicated proposition on the path of economic development, and most low- and middle-income countries and their cities struggle to address the challenges. The impacts of poor waste management are dire and fall disproportionally on the poor, who are often unserved or have little influence on the waste being disposed of formally or informally near their homes.

Waste management data are critical to creating policy and planning for the local context. Understanding how much waste is generated—especially with rapid urbanization and population growth—as well as the types of waste being generated, allows local governments to select appropriate management methods and plan for future demand. This knowledge allows governments to design systems with a suitable number of vehicles, establish efficient routes, set targets for diversion of waste, track progress, and adapt as waste generation patterns change. With accurate data, governments can realistically allocate budget and land, assess relevant technologies, and consider strategic partners, such as the private sector or nongovernmental organizations, for service provision.

This report builds on previous World Bank publications from 2012 and 1999 titled *What a Waste: A Global Review of Solid Waste Management* (Hoornweg and Bhada-Tata 2012) and *What a Waste: Solid Waste Management in Asia* (Hoornweg and Thomas 1999). This current edition of *What a Waste* expands on the type of data collected and includes 217 countries and economies and 367 cities. The data are updated to recent years, and the waste generation data are scaled to a single year to allow for comparison across countries and economies. The projections for waste generation use the most comprehensive database available to date to determine how waste generation dynamically changes based on changes in economic development and population growth. The metrics included in this report expand from solid waste management generation, composition, collection, treatment, and disposal to include information on financing and costs, institutional arrangements and policies, administrative and operational models, citizen engagement, special wastes, and the informal sector.

Although the data from the past and current publications are not fully comparable because of methodological differences, there are some clear trends to report since 2012. The change in the composition of waste in low-income countries reflects changes in consumption patterns—the share of organic waste fell from 64 percent to 56 percent. The collection of waste in low-income countries significantly increased from about 22 percent to 39 percent, reflecting the prioritization of adequate waste collection in cities and countries. This progress is complemented by an overall global trend of increased recycling and composting. Finally, waste-to-energy incineration in upper-middle-income countries markedly increased from 0.1 percent to 10 percent, driven by China's shift to incineration.

What a Waste 2.0: A Global Snapshot of Solid Waste Management to 2050 targets decision makers, policy makers, and influencers globally, including local governments, international organizations, academics, researchers, nongovernmental organizations, civil society, and financiers. The aim of this report is to share objective waste management data and trends, as well as good and unique international practices, with the hope of improving waste management globally and enabling the optimal use of limited resources.

The world generates 2.01 billion tonnes of municipal solid waste[1] annually, with at least 33 percent of that—extremely conservatively—not managed in an environmentally safe manner. Worldwide, waste generated per person per day averages 0.74 kilogram but ranges widely, from 0.11 to 4.54 kilograms. Though they only account for 16 percent of the world's population, high-income countries generate about 34 percent, or 683 million tonnes, of the world's waste.

When looking forward, global waste is expected to grow to 3.40 billion tonnes by 2050. There is generally a positive correlation between waste generation and income level. Daily per capita waste generation in high-income countries is projected to increase by 19 percent by 2050, compared to low- and middle-income countries where it is anticipated to increase by approximately 40 percent or more. Waste generation was generally found to increase at a faster rate for incremental income changes at lower income levels than at high income levels. The total quantity of waste generated in low-income countries is expected to increase by more than three times by 2050. The East Asia and Pacific region is generating most of the world's

Photo 1.1   Plastic Waste at the Thilafushi Waste Disposal Site, Maldives

Photo 1.2   A Recycler Transports Waste Using a Modified Motorcycle, Bangkok, Thailand

waste, at 23 percent, and the Middle East and North Africa region is producing the least in absolute terms, at 6 percent. However, the fastest growing regions are Sub-Saharan Africa, South Asia, and the Middle East and North Africa where, by 2050, total waste generation is expected to nearly triple, double, and double, respectively. In these regions, more than half of waste is currently openly dumped, and the trajectories of waste growth will have vast implications for the environment, health, and prosperity, thus requiring urgent action.

Waste collection is a critical step in managing waste, yet rates vary largely by income levels, with upper-middle- and high-income countries providing nearly universal waste collection. Low-income countries collect about 48 percent of waste in cities, but this proportion drops drastically to 26 percent outside of urban areas. Across regions, Sub-Saharan Africa collects about 44 percent of waste while Europe and Central Asia and North America collect at least 90 percent of waste.

Waste composition differs across income levels, reflecting varied patterns of consumption. High-income countries generate relatively less food and green waste, at 32 percent of total waste, and generate more dry waste that could be recycled, including plastic, paper, cardboard, metal, and glass, which account for 51 percent of waste. Middle- and low-income countries generate 53 percent and 56 percent food and green waste, respectively, with the fraction of organic waste increasing as economic development levels decrease. In low-income countries, materials that could be recycled account for only 16 percent of the waste stream. Across regions, there is not much variety within waste streams beyond those aligned with income. All regions generate about 50 percent or more organic waste, on average, except for Europe and Central Asia and North America, which generate higher portions of dry waste.

It is a frequent misconception that technology is the solution to the problem of unmanaged and increasing waste. Technology is not a panacea and is usually only one factor to consider when managing solid waste. Countries that advance from open dumping and other rudimentary waste management methods are more likely to succeed when they select locally appropriate solutions. Globally, most waste is currently dumped or disposed of in some form of a landfill. Some 37 percent of waste is disposed of in some form of a landfill, 8 percent of which is disposed of in sanitary landfills with landfill gas collection systems. Open dumping accounts for about 33 percent of waste, 19 percent is recovered through recycling and composting, and 11 percent is incinerated for final disposal. Adequate waste disposal or treatment, such as controlled landfills or more stringently operated facilities, is almost exclusively the domain of high- and upper-middle-income countries. Lower-income countries generally rely on open dumping; 93 percent of waste is dumped in low-income countries and only 2 percent in high-income countries. Upper-middle-income countries have the highest percentage of waste in landfills, at 54 percent. This rate decreases in high-income countries to 39 percent, with diversion of 35 percent of waste to recycling and composting and 22 percent to incineration. Incineration is used primarily in high-capacity, high-income, and land-constrained countries.

Based on the volume of waste generated, its composition, and how it is managed, it is estimated that 1.6 billion tonnes of carbon dioxide (CO_2) equivalent greenhouse gas emissions were generated from solid waste treatment and disposal in 2016, driven primarily by open dumping and disposal in landfills without landfill gas capture systems. This is about 5 percent of global emissions.[2] Solid waste–related emissions are anticipated to increase to 2.6 billion tonnes of CO_2-equivalent per year by 2050 if no improvements are made in the sector.

In most countries, solid waste management operations are typically a local responsibility, and nearly 70 percent of countries have established institutions with responsibility for policy development and regulatory oversight in the waste sector. About two-thirds of countries have created targeted legislation and regulations for solid waste management, though enforcement varies drastically. Direct central government involvement in waste service provision, other than regulatory oversight or fiscal transfers, is uncommon, with about 70 percent of waste services being overseen directly by local public entities. At least half of services, from primary waste collection through treatment and disposal, are operated by public entities and about one-third involve a public-private partnership. However, successful partnerships with the private sector for financing and operations tend to succeed only under certain conditions with appropriate incentive structures and enforcement mechanisms, and therefore they are not always the ideal solution.

Financing solid waste management systems is a significant challenge, even more so for ongoing operational costs than for capital investments, and operational costs need to be taken into account upfront. In high-income countries, operating costs for integrated waste management, including collection, transport, treatment, and disposal, generally exceed $100 per tonne. Lower-income countries spend less on waste operations in absolute terms, with costs of about $35 per tonne and sometimes higher, but these countries experience much more difficulty in recovering costs. Waste management is labor intensive and costs of transportation alone are in the range of $20–$50 per tonne. Cost recovery for waste services differs drastically across income levels. User fees range from an average of $35 per year in low-income countries to $170 per year in high-income countries, with full or nearly full cost recovery being largely limited to high-income countries. User fee models may be fixed or variable based on the type of user being billed. Typically, local governments cover about 50 percent of investment costs for waste systems, and the remainder comes mainly from national government subsidies and the private sector.

The solid waste data presented in this report tell the story of global, regional, and urban trends. The book presents analyses and case studies in the following chapters:

- *Chapter 2: At a Glance: A Global Picture of Solid Waste Management.* Chapter 2 provides an overview of global solid waste management trends related to waste generation, composition, collection, and disposal.
- *Chapter 3: Regional Snapshots.* Chapter 3 provides analyses of waste generation, composition, collection, and disposal across seven regions—East Asia and Pacific, Europe and Central Asia, Latin America and the Caribbean, the Middle East and North Africa, South Asia, Sub-Saharan Africa, and North America.
- *Chapter 4: Waste Administration and Operations.* Chapter 4 provides planning, administrative, operational, and contractual models for solid waste management.
- *Chapter 5: Financing and Cost Recovery for Waste Management Systems.* Chapter 5 highlights typical financing methods and cost recovery options that are being implemented globally.
- *Chapter 6: Waste and Society.* Chapter 6 provides insights into how climate change, technology trends, citizens, and the informal sector all interact with and affect the solid waste management sector.
- *Chapter 7: Case Studies.* Chapter 7 details good and unique practices of waste management around the world, from cost recovery to coordination between different levels of government.

Please refer to maps 1.1 and 1.2 for the definitions of regions and income levels used in this report.

Map 1.1   **Definition of Income Levels**

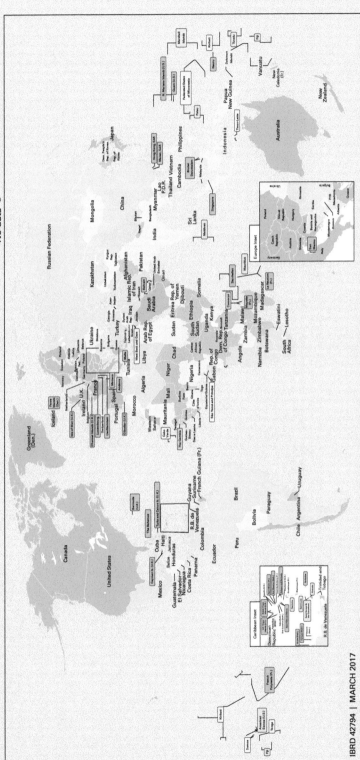

The world by income
- Low ($1,025 or less)
- Lower middle ($1,026–$4,035)
- Upper middle ($4,036–$12,475)
- High ($12,476 or more)
- No data

Classified according to
World Bank estimates of
2015 GNI per capita

IBRD 42794 | MARCH 2017

Note: GNI = gross national income.

Map 1.2   Definition of Regions

Classified according to
World Bank analytical
grouping

- ⬤ East Asia and Pacific
- ⬤ Europe and Central Asia
- ⬤ Latin America and the Caribbean
- ⬤ Middle East and North Africa
- ○ North America
- ⬤ South Asia
- ○ Sub-Saharan Africa
- ○ No data

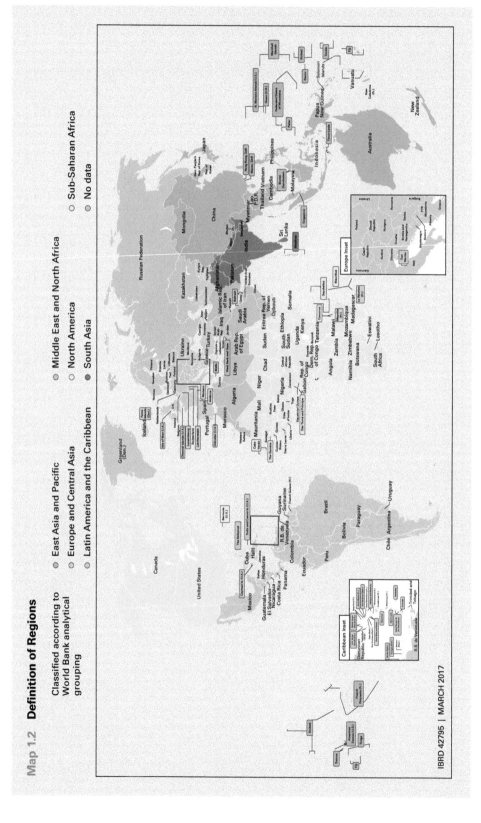

A Note on Data

The *What a Waste* report compiles solid waste management data from various sources and publications and examines the data to provide meaningful trends and analyses for policy makers and researchers. For the purposes of this report, the definition of solid waste encompasses residential, commercial, and institutional waste. Industrial, medical, hazardous, electronic, and construction and demolition waste are reported separately from total national waste generation to the extent possible. Every effort has been undertaken to verify sources and find the most recent information available.

In general, solid waste data should be considered with a degree of caution because of inconsistencies in definitions, data collection methodologies, and availability. The reliability of solid waste data is influenced by several factors, including undefined words or phrases; incomplete or inconsistent definitions; lack of dates, methodologies, or original sources; inconsistent or omitted units; and estimates based on assumptions. Where possible, actual values are presented rather than estimations or projections, even if that requires using older data. In addition, when a source only provides a range for a data point, the average of the range is used for this study and is noted as such. Given the variety of methodologies used by sources, these data are not meant to be used for ordinal ranking of countries or cities but rather to provide trends.

The data reported are predominantly from 2011–17 although overall data span about two decades. Within a single country or city, data availability may cut across several years. Similarly, the year of origin for a specific indicator may vary across countries or cities. The year cited in the tables refers to the year of the data points. However, when a specific year is not available in the original source, the year of the publication is provided instead. Furthermore, when a year range is reported in the original source, the final year of the range is provided in this report's data set.

At a national level, this *What a Waste* study focuses on total waste generation rather than aggregated urban or rural waste generation because of data availability. By providing total waste generation, this study enables comparison across countries, income levels, and regions. To enable cross-comparability of data, all national waste generation statistics are adjusted to a common year using the methodology discussed in box 2.1., with original figures provided in appendix A. However, because urban data are essential for decision making and benchmarking, this study also reports data and trends from 367 cities.

To further maximize cross-comparability of data, statistics for waste composition, collection rates, and disposal methods are consistently reported as percentages in this report. Therefore, data reported by weight or population in the original sources have been converted to percentages wherever possible, and modifications are noted in the comments.

An overview of the methods used for several core indicators is as follows:

Solid Waste Generation

- Data on waste generation at the country level are available for 215 countries and economies.
- Sources reported solid waste data in multiple ways, including total waste generation for the country, daily waste generation rates for the country, and per capita waste generation rates for the whole country or urban areas only.
- In rare cases in which national waste generation data were not available, total waste generation was estimated. Rural solid waste generation rates were estimated to be half that of an aggregate urban rate or that of one or more representative cities. The estimate of one-half as a rural-urban waste generation ratio is supported by several studies and is a conservative estimate that falls below trends observed in available data across regions (Karak, Bhagat, and Bhattacharyya 2012; GIZ and SWEEP-Net, various years). Total waste generation for the whole country was calculated by multiplying waste generation rates by urban and rural populations, using World Bank population data. This methodology mainly applied to 31 countries in the Sub-Saharan Africa region and 8 countries in other regions. The methodology followed for the Sub-Saharan Africa region is explained in box 1.1. Appendix A indicates whether a national waste generation figure was directly reported or was estimated.

Box 1.1 Data for the Sub-Saharan Africa Region

The Sub-Saharan Africa region generates a significant amount of solid waste, and this amount is expected to increase at a higher rate than for any other region given the high rate of urbanization and population growth in the coming decades (Hoornweg and Freire 2013). Although data availability is increasing significantly, statistics on waste generation, collection, treatment, and disposal in the region are currently relatively limited. The data that are available can follow varied definitions, methodologies, and collection methods, and span 23 years from 1993 to 2016.

Given the significance of Sub-Saharan Africa for solid waste generation in the future and the recognition of solid waste management as a priority by many national governments, this report provides estimates for waste generation for many African countries for which country-level data are not available.

To develop data-driven estimates, city-level data were used to extrapolate waste generation to the country level. Out of 48 countries in the Sub-Saharan Africa region, data were available at the country level for 13, or slightly more than a quarter of the total. For 31 countries (about 65 percent), one or more city waste generation rates, typically including the capital city, were used to estimate waste generation for the whole country. The city waste generation rate was used as a proxy rate for the urban population in the country. Half of the urban waste generation rate was used as an estimate for rural waste generation. For the remaining four countries for which no city-level data were available, an average waste generation rate for Africa was used as a proxy for the total amount generated for the country using national population.

- In this report, all figures shown use national waste generation statistics that are adjusted to a common base year of 2016, for cross-comparability. This analysis was conducted using the World Bank's World Development Indicators' gross domestic product (GDP) per capita, PPP data (constant 2011 international $) in conjunction with United Nations population statistics (UN 2017). National waste generation rates for 2016 are estimated using a projection model that is further detailed in box 2.1. All original numbers are provided in appendix A.
- Solid waste generation can be estimated or measured at various places, including at the generation source, point of collection, or disposal site, which may affect the amount of waste reported by sources. This report cites the most reliable measurements available.

Solid Waste Composition

- Waste composition refers to the components of the waste stream as a percentage of the total mass generated.
- In a few cases, composition values do not add up to 100 percent or sum to more than 100 percent when data are cited from multiple sources. Data values provided are as reported in the original source.
- In summary statistics, food, yard, and green waste are combined into one category as food and green waste.

Waste Collection Coverage

- Waste collection coverage data are reported according to multiple definitions: amount of waste collected, number of households served, population served, or geographic area covered. This report analyzes the type of collection coverage reported for countries and cities. If multiple values were reported, the maximum collection rate was used to represent the national or urban collection rate in summary statistics.
- Waste collection coverage is reported at the country level as well as for urban and rural areas, where data are available.

Waste Treatment and Disposal

- Waste treatment and disposal includes recycling, composting, anaerobic digestion, incineration, landfilling, open dumping, and dumping in marine areas or waterways. Given the variability of types of landfills used, data were collected for three types of landfills: sanitary landfills with landfill gas collection systems, controlled landfills that are engineered but for which landfill gas collection systems do not exist or are unknown, and uncategorized landfills. In summary statistics, all landfills are reported together but detailed data are provided in appendix B.

- In cases where disposal and treatment percentages do not add up to 100 percent or where a portion of waste is uncollected, the remaining amount is categorized as waste "unaccounted for." The analyses, figures, and tables in this report assume that waste not accounted for by formal disposal methods, such as landfills or recycling, is dumped. Waste that is disposed of in waterways and that is managed in low- and middle-income countries in "other" manners is also assumed to be dumped. Breakdowns are available in appendix B.

Municipal Waste Management Financials

- Financial data are collected over a range of years, and accounting practices may vary by location.
- Financial data were collected in local currencies when possible, converted to U.S. dollars based on the annual average exchange rate normalized by purchasing power parity, and adjusted to 2011 using the consumer price index to account for potential differences in inflation and to ensure cross-comparability.
- Financial information for solid waste systems was the most scarce among all data categories. When the number of observations was limited, data were aggregated at an income level rather than by regions, and only metrics with substantial geographic diversity were used for summary statistics.

This edition of *What a Waste* features the results of the most extensive combined national and urban solid waste management data collection effort to date. The current data collection and verification effort was designed to revise and enhance a previous effort in 2012 by expanding national and urban data collection, increasing the scope of metrics included, and providing support to decision makers by sharing good practices and trends globally.

Data for this report were collected through a joint effort by regional experts who consulted local specialists and public agencies, sources in diverse languages, and active waste management facilities. Data were gathered from documents published by local and national governments, international organizations, multilateral and bilateral agencies, journals, books, websites, and news agencies. Data collection primarily took place during 2017. Additionally, regional World Bank solid waste experts provided insights beyond the data collected. These assertions are included in the regional snapshots to provide further context for each region but are not attributed to each expert.

The report aggregates extensive solid waste statistics at the national, urban, and rural levels. The current edition estimates and projects waste generation to 2030 and 2050, taking both urban and rural areas into account. Beyond the core data metrics already detailed, the report provides information on waste management costs, revenues, and tariffs;

special wastes; regulations; public communication; administrative and operational models; and the informal sector. In addition to national-level data for 217 countries and economies, a large amount of data were collected at the city level, for about one to two cities per country or economy.

The most up-to-date data can be accessed through the *What a Waste* website at www.worldbank.org/what-a-waste.

Notes

1. This publication defines municipal solid waste as residential, commercial, and institutional waste. Industrial, medical, hazardous, electronic, and construction and demolition waste are reported separately from total national waste generation to the extent possible.
2. Excluding waste-related transportation.

References

GIZ and SWEEP-Net. 2010. "Country Report on the Solid Waste Management in Syria." Deutsche Gesellschaft für Internationale Zusammenarbeit GmbH (GIZ), Bonn; and Regional Solid Waste Exchange of Information and Expertise Network in Mashreq and Maghreb Countries (SWEEP-Net), on behalf of the German Federal Ministry for Economic Cooperation and Development (BMZ), Tunis.

————. 2014a. "Country Report on Solid Waste Management in Algeria." German Corporation for International Cooperation (Deutsche Gesellschaft für Internationale Zusammenarbeit GmbH [GIZ]), Bonn; and Regional Solid Waste Exchange of Information and Expertise Network in Mashreq and Maghreb Countries (SWEEP-Net), on behalf of the German Federal Ministry for Economic Cooperation and Development (Bundesministerium für wirtschaftliche Zusammenarbeit und Entwicklung [BMZ]), Tunis.

————. 2014b. "Country Report on the Solid Waste Management in Egypt." German Corporation for International Cooperation (Deutsche Gesellschaft für Internationale Zusammenarbeit GmbH [GIZ]), Bonn; and Regional Solid Waste Exchange of Information and Expertise Network in Mashreq and Maghreb Countries (SWEEP-Net), on behalf of the German Federal Ministry for Economic Cooperation and Development (Bundesministerium für wirtschaftliche Zusammenarbeit und Entwicklung [BMZ]), Tunis.

————. 2014c. "Country Report on the Solid Waste Management in Jordan." German Corporation for International Cooperation (Deutsche Gesellschaft für Internationale Zusammenarbeit GmbH [GIZ]), Bonn; and Regional Solid Waste Exchange of Information and Expertise

Network in Mashreq and Maghreb Countries (SWEEP-Net), on behalf of the German Federal Ministry for Economic Cooperation and Development (Bundesministerium für wirtschaftliche Zusammenarbeit und Entwicklung [BMZ]), Tunis.

————. 2014d. "Country Report on the Solid Waste Management in Lebanon." German Corporation for International Cooperation (Deutsche Gesellschaft für Internationale Zusammenarbeit GmbH [GIZ]), Bonn; and Regional Solid Waste Exchange of Information and Expertise Network in Mashreq and Maghreb Countries (SWEEP-Net), on behalf of the German Federal Ministry for Economic Cooperation and Development (Bundesministerium für wirtschaftliche Zusammenarbeit und Entwicklung [BMZ]), Tunis.

————. 2014e. "Country Report on the Solid Waste Management in Morocco." German Corporation for International Cooperation (Deutsche Gesellschaft für Internationale Zusammenarbeit GmbH [GIZ]), Bonn; and Regional Solid Waste Exchange of Information and Expertise Network in Mashreq and Maghreb Countries (SWEEP-Net), on behalf of the German Federal Ministry for Economic Cooperation and Development [Bundesministerium für wirtschaftliche Zusammenarbeit und Entwicklung [BMZ]), Tunis.

————. 2014f. "Country Report on the Solid Waste Management in Occupied Palestinian Territories." German Corporation for International Cooperation (Deutsche Gesellschaft für Internationale Zusammenarbeit GmbH [GIZ]), Bonn; and Regional Solid Waste Exchange of Information and Expertise Network in Mashreq and Maghreb Countries (SWEEP-Net), on behalf of the German Federal Ministry for Economic Cooperation and Development (Bundesministerium für wirtschaftliche Zusammenarbeit und Entwicklung [BMZ]), Tunis.

————. 2014g. "Country Report on the Solid Waste Management in Tunisia." German Corporation for International Cooperation (Deutsche Gesellschaft für Internationale Zusammenarbeit GmbH [GIZ]), Bonn; and Regional Solid Waste Exchange of Information and Expertise Network in Mashreq and Maghreb Countries (SWEEP-Net), on behalf of the German Federal Ministry for Economic Cooperation and Development (Bundesministerium für wirtschaftliche Zusammenarbeit und Entwicklung [BMZ]), Tunis.

Hoornweg, Daniel, and Perinaz Bhada-Tata. 2012. *What a Waste: A Global Review of Solid Waste Management.* Washington, DC: World Bank. https://openknowledge.worldbank.org/handle/10986/17388.

Hoornweg, Daniel, and Mila Freire. 2013. *Building Sustainability in an Urbanizing World: A Partnership Report.* Washington, DC: World Bank. https://openknowledge.worldbank.org/handle/10986/18665.

Hoornweg, Daniel, and Laura Thomas. 1999. *What a Waste: Solid Waste Management in Asia.* Washington, DC: World Bank. http://documents .worldbank.org/curated/en/694561468770664233/What-a-waste-solid -waste-management-in-Asia.

Karak, Tanmoy, R. M. Bhagat, and Pradip Bhattacharyya. 2012. "Municipal Solid Waste Generation, Composition, and Management: The World Scenario." *Critical Reviews in Environmental Science and Technology* 42 (15): 1509–630.

United Nations. 2017. "UN World Population Prospects, Medium Variant Scenario, 2017 Revision. File Name: Total Population—Both Sexes." Department of Economic and Social Affairs, Population Division, United Nations, New York. https://esa.un.org/unpd/wpp/Download/Standard /Population/.

CHAPTER 2

At a Glance: A Global Picture of Solid Waste Management

Key Insights

- The world generates 0.74 kilogram of waste per capita per day, yet national waste generation rates fluctuate widely from 0.11 to 4.54 kilograms per capita per day. Waste generation volumes are generally correlated with income levels and urbanization rates.
- An estimated 2.01 billion tonnes of municipal solid waste were generated in 2016, and this number is expected to grow to 3.40 billion tonnes by 2050 under a business-as-usual scenario.
- The total quantity of waste generated in low-income countries is expected to increase by more than three times by 2050. Currently, the East Asia and Pacific region is generating most of the world's waste, at 23 percent, and the Middle East and North Africa region is producing the least in absolute terms, at 6 percent. However, waste is growing the fastest in Sub-Saharan Africa, South Asia, and the Middle East North Africa regions, where, by 2050, total waste generated is expected to approximately triple, double, and double, respectively.
- Food and green waste comprise more than 50 percent of waste in low- and middle-income countries. In high-income countries the amount of organic waste is comparable in absolute terms but, because of larger amounts of packaging waste and other nonorganic waste, the fraction of organics is about 32 percent.
- Recyclables make up a substantial fraction of waste streams, ranging from 16 percent paper, cardboard, plastic, metal, and glass in low-income countries to about 50 percent in high-income countries. As countries rise in income level, the quantity of recyclables in the waste stream increases, with paper increasing most significantly.

- More than one-third of waste in high-income countries is recovered through recycling and composting.
- Waste collection rates vary widely by income levels. High- and upper-middle-income countries typically provide universal waste collection. Low-income countries tend to collect about 48 percent of waste in cities, but outside of urban areas waste collection coverage is about 26 percent. In middle-income countries, rural waste collection coverage varies from 33 percent to 45 percent.
- Globally, about 37 percent of waste is disposed of in some type of landfill, 33 percent is openly dumped, 19 percent undergoes materials recovery through recycling and composting, and 11 percent is treated through modern incineration.
- Adequate waste disposal or treatment using controlled landfills or more stringently operated facilities is almost exclusively the domain of high- and upper-middle-income countries. Lower-income countries generally rely on open dumping—93 percent of waste is dumped in low-income countries and only 2 percent in high-income countries.
- Upper-middle-income countries practice the highest percentage of land-filling, at 54 percent. This rate decreases in high-income countries to 39 percent, where 35 percent of waste is diverted to recycling and composting and 22 percent to incineration.

Waste Generation

Waste generation is a natural product of urbanization, economic development, and population growth. As nations and cities become more populated and prosperous, offer more products and services to citizens, and participate in global trade and exchange, they face corresponding amounts of waste to manage through treatment and disposal (map 2.1).

The 2012 edition of *What a Waste: A Global Review of Solid Waste Management* estimated global waste production to be 1.3 billion tonnes per year based on available data (Hoornweg and Bhada-Tata 2012). In recent years, waste production has grown at levels consistent with initial projections, and data tracking and reporting have improved substantially. Based on the latest data available, global waste generation in 2016 was estimated to have reached 2.01 billion tonnes.

Countries in the East Asia and Pacific and the Europe and Central Asia regions account for 43 percent of the world's waste by magnitude (figure 2.1, panel a). The Middle East and North Africa and Sub-Saharan Africa regions produce the least amount of waste, together accounting for 15 percent of the world's waste. East Asia and Pacific generates the most in absolute terms, an estimated 468 million tonnes in 2016, and the Middle East and North Africa region generates the least, at 129 million tonnes (figure 2.1, panel b).

Map 2.1   Waste Generation Per Capita

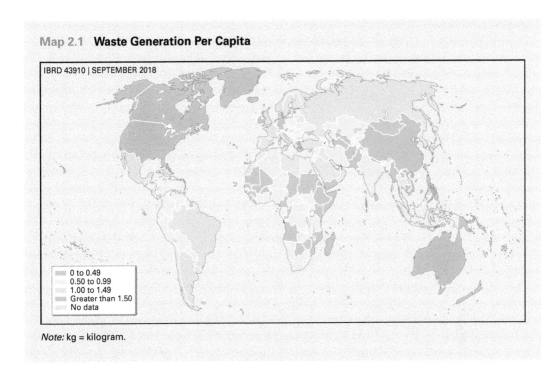

Note: kg = kilogram.

Figure 2.1   Waste Generation by Region

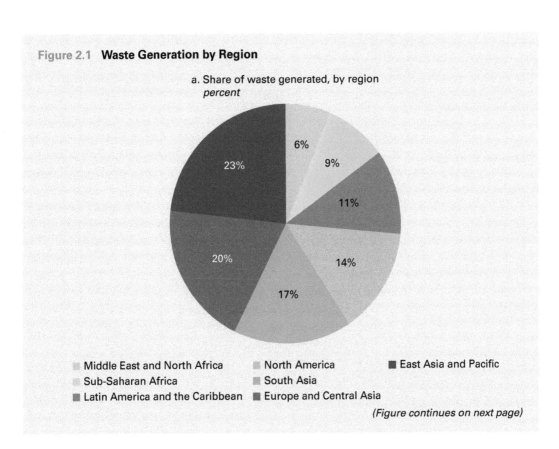

a. Share of waste generated, by region
percent

Middle East and North Africa

Sub-Saharan Africa

Latin America and the Caribbean

North America

South Asia

Europe and Central Asia

East Asia and Pacific

(Figure continues on next page)

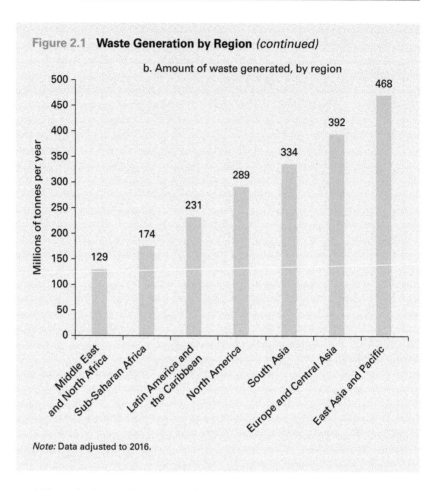

Figure 2.1   **Waste Generation by Region** *(continued)*

b. Amount of waste generated, by region

Note: Data adjusted to 2016.

Although they only account for 16 percent of the world's population, high-income countries generate 34 percent, or 683 million tonnes, of the world's waste (figure 2.2). Low-income countries account for 9 percent of the world's population but generate only about 5 percent of global waste, or 93 million tonnes.

The three countries in the North America region—Bermuda, Canada, and the United States—produce the highest average amount of waste per capita, at 2.21 kilograms per day. All three countries are high-income nations. The three regions with a high proportion of low- and middle-income nations generate the lowest amount of waste per capita: Sub-Saharan Africa averages 0.46 kilogram per day, South Asia 0.52 kilogram per day, and East Asia and Pacific 0.56 kilogram per day. Overall, the estimated global average for 2016 is 0.74 kilogram of waste per capita per day and total generation of solid waste is about 2.01 billion tonnes.

Average waste generation across countries varies substantially, from 0.11 kilogram per capita per day to 4.54 kilograms per capita per day (table 2.1).

Waste generation has an overall positive relationship with economic development (figure 2.3). For incremental income changes, waste generation is generally shown to increase at a faster rate at lower income levels

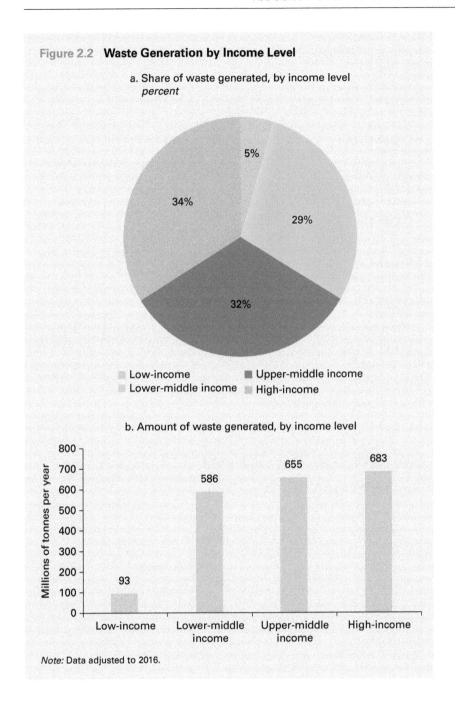

Figure 2.2 **Waste Generation by Income Level**

a. Share of waste generated, by income level
percent

- Low-income
- Lower-middle income
- Upper-middle income
- High-income

b. Amount of waste generated, by income level

Note: Data adjusted to 2016.

than at higher income levels. However, at the lowest income levels, waste generation per capita declines initially with income growth. The slower growth of waste generation at higher income levels could be due to reduced marginal demand for consumption, and therefore reduced waste.

Waste generation also increases with urbanization (figure 2.4). High-income countries and economies are more urbanized and they generate more waste per capita and in total. At a regional level, North America, with

Table 2.1 Ranges of Average National Waste Generation by Region

kg/capita/day

	2016 Average	Min	25th Percentile	75th Percentile	Max
Sub-Saharan Africa	0.46	0.11	0.35	0.55	1.57
East Asia and Pacific	0.56	0.14	0.45	1.36	3.72
South Asia	0.52	0.17	0.32	0.54	1.44
Middle East and North Africa	0.81	0.44	0.66	1.40	1.83
Latin America and Caribbean	0.99	0.41	0.76	1.39	4.46
Europe and Central Asia	1.18	0.27	0.94	1.53	4.45
North America	2.21	1.94	2.09	3.39	4.54

Note: kg = kilogram.

Figure 2.3 Waste Generation and Gross Domestic Product

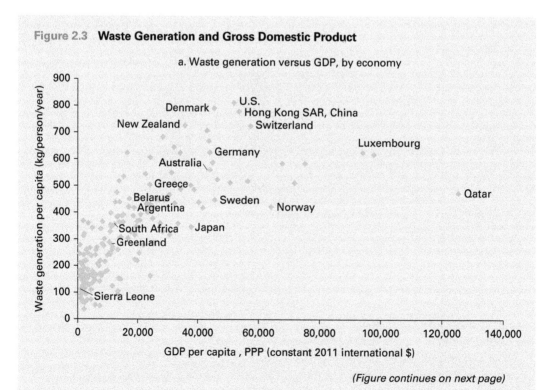

a. Waste generation versus GDP, by economy

(Figure continues on next page)

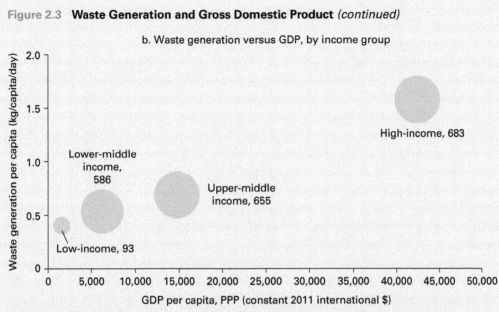

Figure 2.3   **Waste Generation and Gross Domestic Product** *(continued)*

b. Waste generation versus GDP, by income group

Note: Data in panel a are from originally reported year. Data in panel b are adjusted to 2016. Size of bubble in panel b denotes total waste generated in millions of tonnes annually. Waste generation per capita per day: Low income = 0.43 kg, lower-middle income = 0.61 kg, upper-middle income = 0.69 kg, high income = 1.57 kg. GDP = gross domestic product; kg = kilogram.

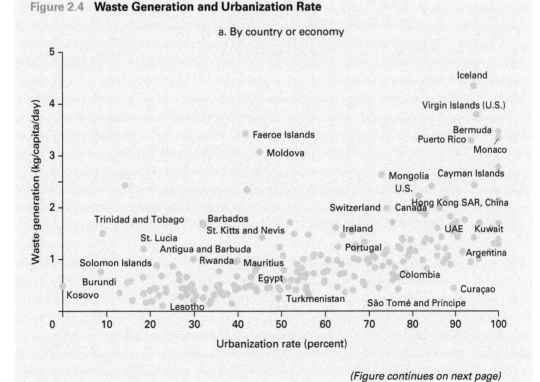

Figure 2.4   **Waste Generation and Urbanization Rate**

a. By country or economy

(Figure continues on next page)

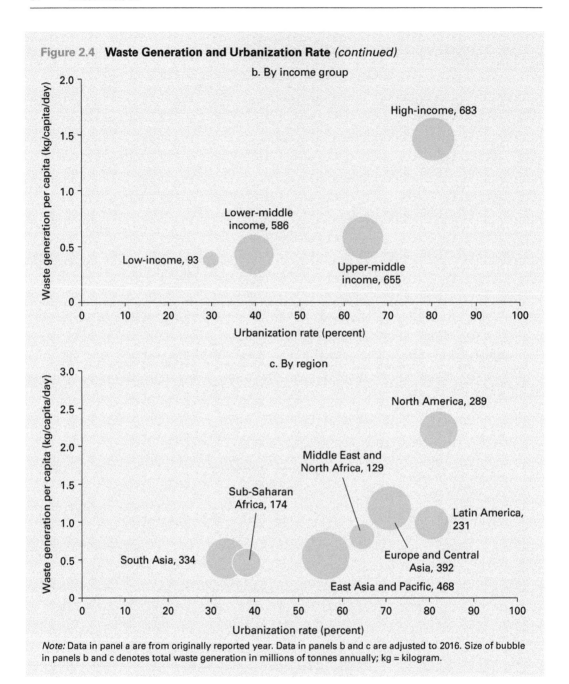

Figure 2.4 **Waste Generation and Urbanization Rate** *(continued)*

b. By income group

Note: Data in panel a are from originally reported year. Data in panels b and c are adjusted to 2016. Size of bubble in panels b and c denotes total waste generation in millions of tonnes annually; kg = kilogram.

the highest urbanization rate at 82 percent, generates 2.21 kilograms per capita per day, while Sub-Saharan Africa generates 0.46 kilogram per capita per day at a 38 percent urbanization rate.

Projected Waste Generation

By 2030, the world is expected to generate 2.59 billion tonnes of waste annually (figure 2.5). By 2050, waste generation across the world is expected to reach 3.40 billion tonnes (see methodology in box 2.1).

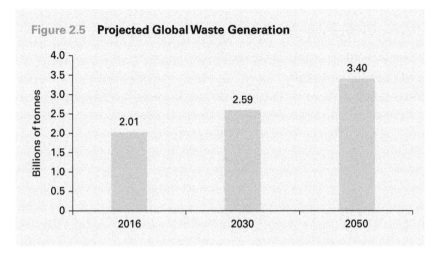

Figure 2.5   **Projected Global Waste Generation**

Box 2.1   Waste Generation Projection Methodology

To ensure cross-comparability of waste generation data and to develop projections for global waste generation, available waste generation data were adjusted from a variety of origin years to 2016, 2030, and 2050.

Key Assumptions

This analysis assumes that waste generation grows primarily based on two factors:

- **Gross domestic product (GDP) growth:** As a country advances economically, its per capita waste generation rates increase. Economic growth is reflected using GDP per capita, with a purchasing power parity adjustment to 2011 to allow for comparison across countries.
- **Population growth:** As a country's population grows, amounts of total waste generated rise accordingly.

Methodology Overview

The model uses the World Bank's World Development Indicator's GDP per capita, PPP (constant 2011 international $) for the waste per capita regression model, the Organisation for Economic Co-operation and Development (OECD) GDP per capita projections, PPP (constant 2005 international $) for the waste per capita projection estimates, and the United Nations (UN) population growth rates to calculate future waste production:

- **Relationship between GDP growth and waste generation rates:** The observed relationship between GDP growth and waste generation is reflected in figure B2.1.1. A regression model was used to capture the relationship between GDP per capita and waste generation per capita. The model was developed using country-level baseline waste generation data from the data collected and GDP per capita data from the associated year. In the model of best fit, the natural logarithm of GDP per capita is the independent variable and tonnes of waste generation per capita is the dependent variable.
- **Proxy waste generation rates:** The regression model was used to estimate the expected growth in each country's waste generation rate based on the growth in that country's GDP per capita. Using the regression model coefficient and intercept, as well as GDP per capita data for the

(Box continues on next page)

Box 2.1   Waste Generation Projection Methodology *(continued)*

Figure B2.1.1   Waste Generation: Actual and Model Prediction

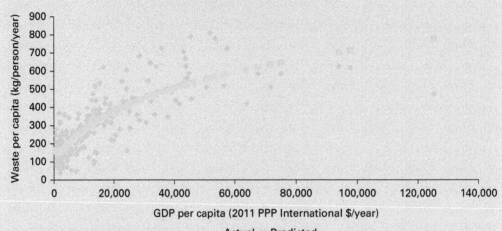

Note: GDP = gross domestic product.

base year and for the projection years, proxy waste generation rates per year were modeled for each country for the base and target years, per equation B2.1.1.

Proxy waste generation per capita
$$= 1647.41 - 419.73 \ln(GDP\ per\ capita) + 29.43 \ln(GDP\ per\ capita)^2 \qquad \text{(B2.1.1)}$$

- **Projected waste generation:** The change in proxy waste generation rates developed through the model was used as the growth rate for waste generation for that country. This growth rate was applied to the actual baseline waste generation per capita rate from the data collected to adjust actual waste generation rates from the base year to 2016, 2030, and 2035, per equation B2.1.2. If a growth rate could not be calculated for an economy or territory because of a lack of GDP data, a regional average was used.

Projected Waste Generation Rate $_{Target Year}$ = *(Proxy Waste Generation Rate* $_{Target Year}$ /
Proxy Waste Generation Rate $_{Base Year}$ *)* ×
Actual Waste Generation Rate $_{Base Year}$ (B2.1.2)

- **2016 waste generation:** The adjusted per capita waste generation rate for 2016 was multiplied by the historical population level for 2016. If waste generation data were already reported for 2016, the original data were used.
- **2030 and 2050 waste generation:** The adjusted per capita waste generation rates for 2030 and 2050 were multiplied by the respective projected population levels for the target year.

In adjusting and projecting waste generation, urbanization rates and changes in country income classification are not considered.

Data Sources
- Waste Generation: Best available national waste generation data from current study
- Base Year and 2016 Population: World Bank Open Data
- 2030 and 2050 Population: UN Population Projections, Medium Variant, 2017 Revision
- GDP per Capita, PPP (constant 2011 international $): World Bank's World Development Indicators
- GDP per Capita, PPP (constant 2005 international $): OECD

High-income countries are expected to experience the least amount of waste generation growth by 2030, given that they have reached a point of economic development at which materials consumption is less linked to gross domestic product growth (figure 2.6, panel a).[1] Low-income countries are positioned for the greatest amount of growth in economic activity as well as population, and waste levels are expected to more than triple by 2050. At a per capita level, trends are similar in that the largest growth in waste generation is expected in low and middle-income countries (figure 2.6, panel b).

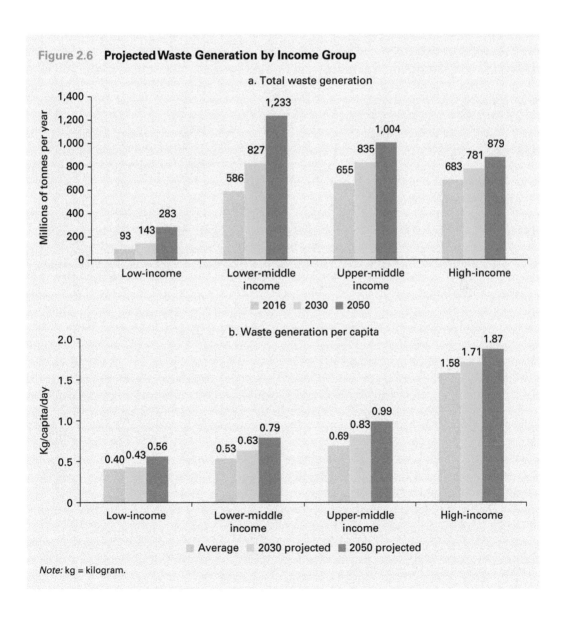

Figure 2.6 **Projected Waste Generation by Income Group**

Note: kg = kilogram.

Since waste generation is generally expected to increase with economic development and population growth, regions with high proportions of growing low-income and lower-middle-income countries are anticipated to experience the greatest increase in waste production. In particular, the Sub-Saharan Africa and South Asia regions are expected to see waste levels approximately triple and double, respectively, in the next three decades with economic growth and urbanization (figure 2.7). Regions with higher-income countries, such as North America and Europe and Central Asia, are expected to see waste levels rise more gradually.

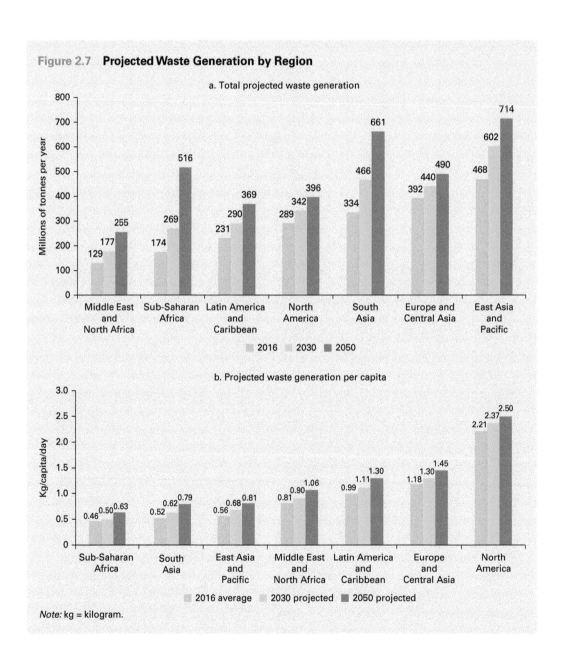

Figure 2.7   Projected Waste Generation by Region

Note: kg = kilogram.

Waste Composition

Waste composition is the categorization of types of materials in municipal solid waste. Waste composition is generally determined through a standard waste audit, in which samples of garbage are taken from generators or final disposal sites, sorted into predefined categories, and weighed.

At an international level, the largest waste category is food and green waste, making up 44 percent of global waste (figure 2.8). Dry recyclables (plastic, paper and cardboard, metal, and glass) amount to another 38 percent of waste.

Waste composition varies considerably by income level (figure 2.9). The percentage of organic matter in waste decreases as income levels rise. Consumed goods in higher-income countries include more materials such as paper and plastic than they do in lower-income countries. The granularity of data for waste composition, such as detailed accounts of rubber and wood waste, also increases by income level.

Global food loss and waste accounts for a significant proportion of food and green waste and is discussed further in box 2.2.

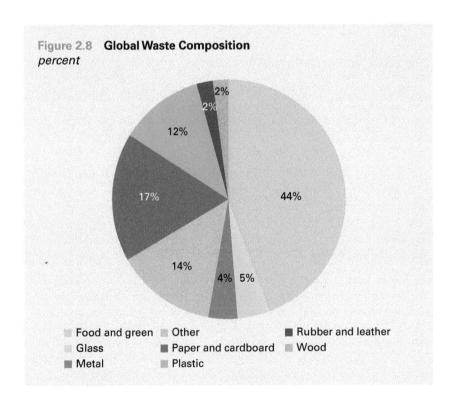

Figure 2.8   Global Waste Composition
percent

Food and green   Other   Rubber and leather
Glass   Paper and cardboard   Wood
Metal   Plastic

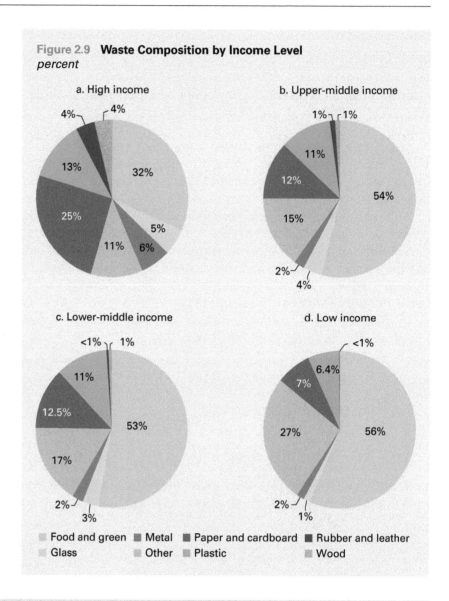

Figure 2.9   **Waste Composition by Income Level**
percent

a. High income

4% — 4%
4%
13%   32%
25%
5%
11%   6%

b. Upper-middle income

1% — 1%
11%
12%   54%
15%
2% —
4%

c. Lower-middle income

<1% — 1%
11%
12.5%   53%
17%
2% —
3%

d. Low income

<1%
6.4%
7%
27%   56%
2% —
1%

Food and green   Metal   Paper and cardboard   Rubber and leather
Glass   Other   Plastic   Wood

Box 2.2   **Global Food Loss and Waste**

Across global food systems, food loss and waste (FLW) is a widespread issue, posing a challenge to food security, food safety, the economy, and environmental sustainability. No accurate estimates of the extent of FLW are available, but studies indicate that FLW is roughly 30 percent of all food globally (FAO 2015). This amounts to 1.3 billion tonnes per year. FLW represents wastage of resources, including the land, water, labor, and energy used to produce food. It strongly contributes to climate change because greenhouse gases are emitted during food production and distribution activities, and methane is released during the decay of wasted food. FLW also affects food supply chains by lowering income for food producers, increasing costs for food

(Box continues on next page)

Box 2.2 **Global Food Loss and Waste** *(continued)*

consumers, and reducing access to food. Minimizing FLW could lead to substantial food security and environmental gains.

The causes of FLW vary across the world and depend on specific local conditions. Typically, FLW in low-income countries occurs at the production, postharvest handling, storage, and processing stages and is caused predominantly by managerial and technical limitations. FLW mostly occurs in the distribution and consumption stages in middle- and high-income countries, although it can happen in earlier stages such as when agricultural subsidies lead to overproduction of farm crops. These challenges relate to consumer behavior and government policies and regulation.

Improving coordination among actors along the different stages of the supply chain could address some of the FLW issues globally. Measures to reduce FLW in low-income countries could involve investment in infrastructure and transportation, including in technology for storage and cooling. Small-scale farmers could also be supported by provision of improved financing and credit to allow them to diversify or scale their production. In high-income countries, consumer education for behavior change is key to decreasing FLW. In addition to decreasing FLW along the supply chain, discarded food could also be managed productively for composting and energy recovery.

Regional and international stakeholders are taking action to address FLW. The African Union is working with 14 governments to translate the "Malabo Declaration on Accelerated Agricultural Growth and Transformation for Shared Prosperity and Improved Livelihoods," including food loss reduction, into proper national policy and strategies in Africa (African Union Commission 2014). The Deputy-Secretary General of the United Nations called on all partners to adopt a more holistic approach to food security, one that prioritizes FLW, builds new coalitions, scales up current work, and innovates (Helvetas 2018). The Food and Agriculture Organization has been working on developing new metrics and methodologies to measure FLW, and the organization's SAVE FOOD Initiative works with civil society to address the issue (FAO 2018). The World Food Programme is including food loss as part of some five-year country plans in Africa and launched the Farm to Market Alliance to structure local markets and promote loss reduction technologies among smallholder farmers (World Food Program 2017). The World Bank is tackling the issue through loans, such as in Argentina, and by coordinated food waste management and the establishment of a cross-sectoral strategy (World Bank 2015).

Several national and local governments have also taken action. In 2016, the government of Italy approved a law to enhance collaboration among key stakeholders, educate the public, encourage food donations from business through financial incentives, and promote reusable and recyclable packaging (Azzuro, Gaiani, and Vittuari 2016). In 2016, France became the first country in the world to ban supermarkets from throwing away or destroying unsold food, forcing them instead to donate it to charities and food banks (Chrisafis 2016). In 2009, the city of San Francisco in the United States passed an ordinance requiring all residents and tourists to compost food waste (McClellan 2017). The city of Ningbo in China diverts food waste from apartment buildings to an anaerobic digestion facility (Lee et al. 2014). In several cities in Sweden, biogas is produced from food waste to power vehicles and generate heat (Swedish Gas Centre, Swedish Gas Association, and Swedish Biogas Association 2008). In cities like Linköping, Sweden, the majority of public buses have been converted to use recovered biogas. The optimal strategy to reduce loss and recover food waste depends greatly on the local context, but the increasing global action reveals the many policy, technology, and educational avenues available.

Waste Collection

Waste collection is one of the most common services provided at a municipal level. Several waste collection service models are used across the globe. The most common form of waste collection is door-to-door collection. In this model, trucks or small vehicles—or, where environments are more constrained, handcarts or donkeys—are used to pick up garbage outside of households at a predetermined frequency. In certain localities, communities may dispose of waste in a central container or collection point where it is picked up by the municipality and transported to final disposal sites. In other areas with less regular collection, communities may be notified through a bell or other signal that a collection vehicle has arrived in the neighborhood, such as in Taiwan, China (see case study 15 in chapter 7).

Waste collection rates in high-income countries and in North America are near 100 percent (figure 2.10).[2] In lower-middle-income countries, collection rates are about 51 percent, and in low-income countries, about 39 percent. In low-income countries, uncollected waste is often managed independently by households and may be openly dumped, burned, or, less commonly, composted. Improvement of waste collection services is a critical step to reduce pollution and thereby to improve human health and, potentially, traffic congestion.

Waste collection rates tend to be substantially higher for urban areas than for rural areas, since waste management is typically an urban service. In lower-middle-income countries, waste collection rates are more than twice as high in cities as in rural areas (figure 2.11).

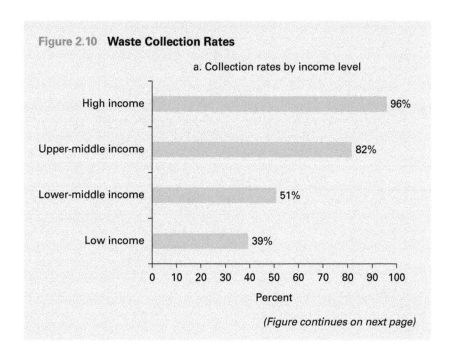

Figure 2.10 **Waste Collection Rates**

a. Collection rates by income level

Income level	Percent
High income	96%
Upper-middle income	82%
Lower-middle income	51%
Low income	39%

(Figure continues on next page)

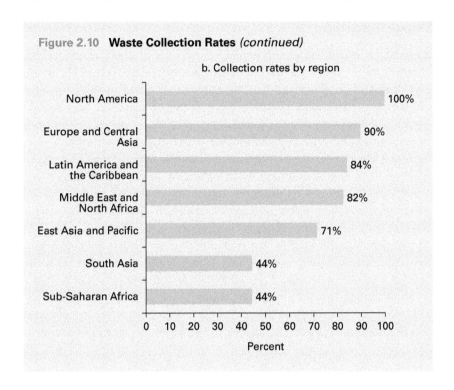

Figure 2.10 **Waste Collection Rates** *(continued)*

b. Collection rates by region

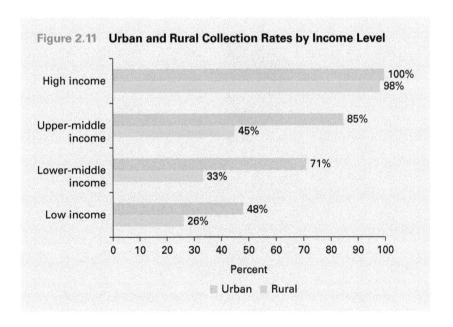

Figure 2.11 **Urban and Rural Collection Rates by Income Level**

Waste Disposal

Around the world, almost 40 percent of waste is disposed of in landfills (figure 2.12).[3] About 19 percent undergoes materials recovery through recycling and composting,[4] and 11 percent is treated through modern incineration. Although globally 33 percent of waste is still openly dumped,[5] governments are increasingly recognizing the risks and costs of dumpsites and pursuing sustainable waste disposal methods.

Waste disposal practices vary significantly by income level and region (figure 2.13). Open dumping is prevalent in lower-income countries, where landfills are not yet available. About 93 percent of waste is burned or dumped in roads, open land, or waterways in low-income countries, whereas only 2 percent of waste is dumped in high-income countries. More than two-thirds of waste is dumped in the South Asia and Sub-Saharan Africa regions, which will significantly impact future waste growth.

As nations prosper economically, waste is managed using more sustainable methods. Construction and use of landfills is commonly the first step toward sustainable waste management. Whereas only 3 percent of waste is

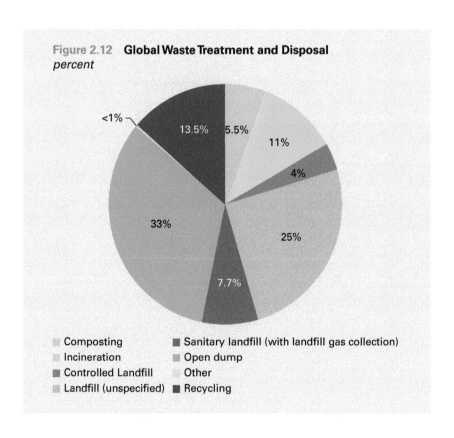

Figure 2.12 **Global Waste Treatment and Disposal**
percent

<1%
13.5%
5.5%
11%
4%
25%
7.7%
33%

Composting   ■ Sanitary landfill (with landfill gas collection)
Incineration   Open dump
■ Controlled Landfill   Other
Landfill (unspecified)   ■ Recycling

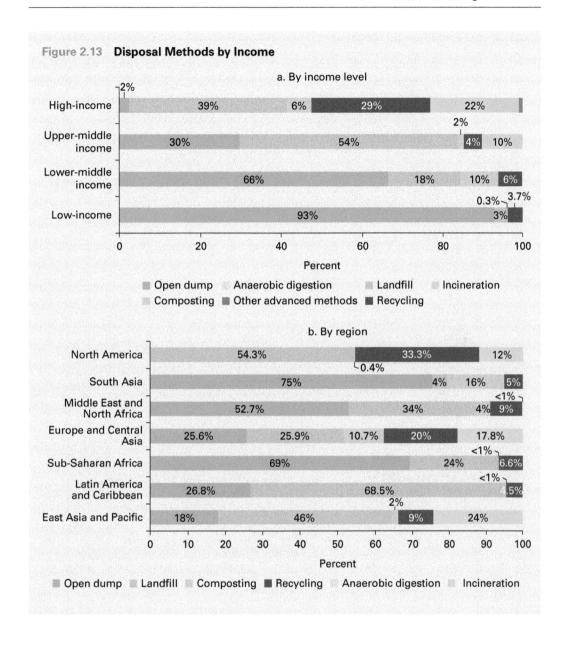

Figure 2.13   Disposal Methods by Income

a. By income level

Legend: Open dump · Anaerobic digestion · Landfill · Incineration · Composting · Other advanced methods · Recycling

High-income: 2%, 39%, 6%, 29%, 22%
Upper-middle income: 30%, 54%, 2%, 4%, 10%
Lower-middle income: 66%, 18%, 10%, 6%
Low-income: 93%, 3%, 0.3%, 3.7%

b. By region

Legend: Open dump · Landfill · Composting · Recycling · Anaerobic digestion · Incineration

North America: 54.3%, 33.3%, 12%, 0.4%
South Asia: 75%, 4%, 16%, 5%
Middle East and North Africa: 52.7%, 34%, 4%, 9%, <1%
Europe and Central Asia: 25.6%, 25.9%, 10.7%, 20%, 17.8%
Sub-Saharan Africa: 69%, 24%, 6.6%, <1%
Latin America and Caribbean: 26.8%, 68.5%, 4.5%, <1%
East Asia and Pacific: 18%, 46%, 2%, 9%, 24%

deposited in landfills in low-income countries, about 54 percent of waste is sent to landfills in upper-middle-income countries. Furthermore, wealthier countries tend to put greater focus on materials recovery through recycling and composting. In high-income countries, 29 percent of waste is recycled and 6 percent composted. Incineration is also more common. In high-income countries, 22 percent of waste is incinerated, largely within high-capacity and land-constrained countries and territories such as Japan and the British Virgin Islands.

Special Wastes

Municipal solid waste is one of several waste streams that countries and cities manage. Other common waste streams include industrial waste, agricultural waste, construction and demolition waste, hazardous waste, medical waste, and electronic waste, or e-waste (figure 2.14).

Some waste streams, such as industrial waste, are generated in much higher quantities than municipal solid waste (table 2.2). For the countries with

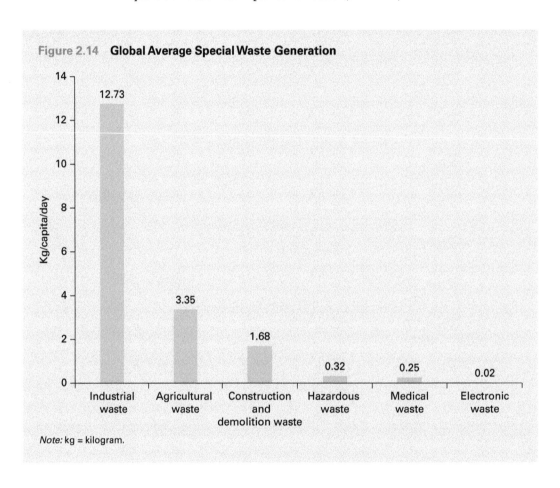

Figure 2.14   Global Average Special Waste Generation

Note: kg = kilogram.

Table 2.2   Industrial and Electronic Waste Generation Rates
kg/capita/day

	Industrial waste generation	E-waste generation
High income	42.62	0.05
Upper-middle income	5.72	0.02
Lower-middle income	0.36	0.01
Low income	No data	<0.01

Note: kg = kilogram.

available industrial waste generation data, the trend shows that globally, industrial waste generation is almost 18 times greater than municipal solid waste. Generation of industrial waste rises significantly as income level increases.

Global agricultural waste production is more than four and a half times that of municipal solid waste. Agricultural waste is most significant in countries with large farming industries. Agricultural waste is often managed separately from other waste streams since it is largely organic and may serve as a useful input for future agricultural activities.

Construction and demolition waste may compete with municipal solid waste for disposal space in landfills. In some countries, such as India, it is common to dispose of both in the same disposal facilities.

Hazardous, medical, and e-waste are typically only a fraction of municipal solid waste. If disposed of properly, these wastes are typically treated in specialized facilities, including chemical processing plants, incinerators, and disassembly centers, respectively. The generation of e-waste is associated with economic development, with high-income countries generating five times the volume of e-waste generated by lower-middle-income countries. The increasing amount of e-waste and its potential for environmental pollution and recycling may be an area of consideration for rapidly developing countries.

Notes

1. Income levels for countries are held constant to classifications at the time of publication; that is, potential changes in income level category are not considered for total projected waste generation levels.
2. The collection rate for North America is rounded from 99.7 percent.
3. Landfills may be controlled, sanitary, or unspecified.
4. Some countries report waste that is composted under the category "Recycling." These two disposal methods may be viewed together as materials recovery.
5. Waste that is uncollected, classified as treated by "Other" methods, thrown in waterways, and not accounted for by any disposal method is considered dumped. "Other" typically refers to the open burning of waste.

References

African Union Commission. 2014. "Malabo Declaration on Accelerated Agricultural Growth and Transformation for Shared Prosperity and Improved Livelihoods." African Union Commission, Addis Ababa, Ethiopia. http://www.resakss.org/sites/default/files/Malabo%20Declaration%20on%20Agriculture_2014_11%2026-.pdf.

Azzuro, Paolo, Silvia Gaiani, and Matteo Vittuari. 2016. "Italy–Country Report on National Food Waste Policy." Fusions EU, Bologna. http://www.eu-fusions.org/phocadownload/country-report/FUSIONS%20IT%20Country%20Report%2030.06.pdf.

Chrisafis, Angelique. 2016. "French Law Forbids Food Waste by Supermarkets." *The Guardian*, February 4. https://www.theguardian.com /world/2016/feb/04/french-law-forbids-food-waste-by-supermarkets.

FAO (Food and Agriculture Organization). 2015. *Global Initiative on Food Loss and Waste Reduction*. Rome: Food and Agriculture Organization of the United Nations. http://www.fao.org/3/a-i4068e.pdf.

———. 2018. "SAVE FOOD: Global Initiative on Food Loss and Waste Reduction." Food and Agriculture Organization of the United Nations, Rome. http://www.fao.org/save-food/news-and-multimedia/news/news -details/en/c/1105834/.

Helvetas. 2018. "Improved Postharvest Management." https://www .helvetas.org/en/albania/how-you-can-help/follow-us/blog/other-topics /Improved-postharvest-management.

Hoornweg, Daniel, and Perinaz Bhada-Tata. 2012. *What a Waste: A Global Review of Solid Waste Management*. Washington, DC: World Bank. https://openknowledge.worldbank.org/handle/10986/17388.

Lee, Marcus, Farouk Banna, Renee Ho, Perinaz Bhada-Tata, and Silpa Kaza. 2014. *Results-Based Financing for Municipal Solid Waste*. Washington, DC: World Bank.

McClellan, Jennifer. 2017. "How San Francisco's Mandatory Composting Laws Turn Food Waste into Profit." azcentral. https://www.azcentral .com/story/entertainment/dining/food-waste/2017/08/03/san-francisco -mandatory-composting-law-turns-food-waste-money/440879001/.

OECD (Organisation for Economic Co-operation and Development). Forthcoming. "Projections of Materials Use to 2060 and Their Economic Drivers." OECD Publishing, Paris.

Swedish Gas Centre, Swedish Gas Association, and Swedish Biogas Association. 2008. Handbook. Stockholm. http://www.greengasgrids.eu /fileadmin/greengas/media/Markets/Sweden/BiogasinfoEngGoda Exempel.pdf.

United Nations. 2017. *World Population Prospects: The 2017 Revision, Key Findings and Advance Tables*. Working Paper No. ESA/P/WP/248. Medium Variant Scenario. New York: Department of Economic and Social Affairs, Population Division, United Nations.

World Bank. 2015. "Eyes Bigger than Belly: A Habit Which Is Harming Latin America." World Bank, Washington, DC. http://www.worldbank .org/en/news/feature/2015/11/12/food-waste-habit-harming -latin-america.

World Bank Open Data: Population 1960–2016. n.d. https://data.world bank.org/indicator/SP.POP.TOTL.

World Food Programme. 2017. "Farm to Market Alliance." World Food Program, Rome. https://innovation.wfp.org/project/farm-market-alliance.

CHAPTER 3

Regional Snapshots

East Asia and Pacific

Key Insights

- The East Asia and Pacific region generated the most waste globally at 468 million tonnes in 2016, an average of 0.56 kilogram per capita per day.
- Some 53 percent of waste in East Asia and Pacific is composed of food and green waste, and dry recyclables comprise about one-third of the waste.
- Waste collection coverage in East Asia and Pacific is about 77 percent at an urban level, 45 percent at a rural level, and 71 percent overall.
- About 46 percent of garbage is disposed of in some form of landfill; 24 percent of waste is incinerated, mainly in high-income countries; and about 9 percent of waste is recycled.
- Cities are increasingly developing source-separation and recycling programs for both dry materials and organics.

Background and Trends

The East Asia and Pacific region consists of 37 countries and economies on the main Asian continent, Australia, and surrounding island states in the Pacific Ocean. The region was home to a population of 2.27 billion in 2016. Disposal practices vary in East Asia and Pacific. Although open dumping remains a common disposal practice, higher-income countries such as China and the Republic of Korea have achieved high landfilling and recycling rates. Because of increasingly rigorous environmental laws, disposal practices are making the transition to sanitary landfills. Island states in the Pacific are especially focused on materials recovery through recycling and composting.

From an administrative perspective, waste systems are increasingly becoming privatized in municipalities, and cities are developing structures for accountability and quality. Governing agencies are exploring ways to reduce overlaps in responsibility. Financially, waste systems are heavily subsidized by the government, and countries with unplanned settlements, such as Mongolia, still experience difficulty recovering costs. However, fee-recovery systems are maturing. For example, several countries charge waste management user fees through combined utility bills, and the use of behavior-changing variable fees is famously practiced in Korea, as well as in several other countries, including the Philippines and Thailand.

Waste Generation and Composition

The East Asia and Pacific region generated 468 million tonnes of waste in 2016, at an average rate of 0.56 kilogram per person per day (figure 3.1). The largest waste generators are typically high-income countries or island states. About 47 percent of waste in the region is generated by the economic hub of China, which is home to 61 percent of the region's population. However, at 0.43 kilogram, China's daily per capita waste generation rate is below the regional average, reflecting the lower amounts of waste generated by the country's significant rural population.

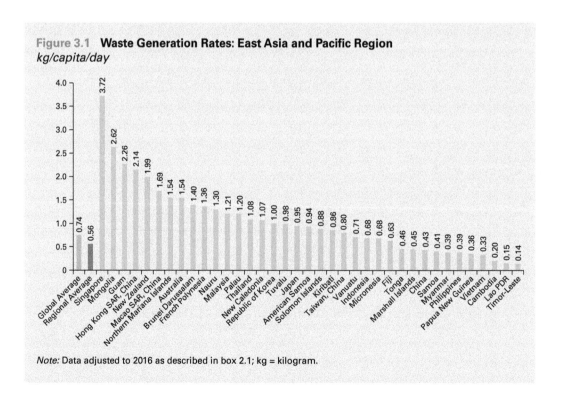

Figure 3.1 Waste Generation Rates: East Asia and Pacific Region
kg/capita/day

Note: Data adjusted to 2016 as described in box 2.1; kg = kilogram.

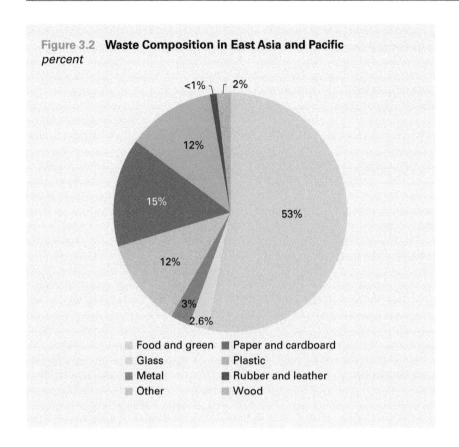

Figure 3.2   **Waste Composition in East Asia and Pacific**
percent

- Food and green
- Glass
- Metal
- Other
- Paper and cardboard
- Plastic
- Rubber and leather
- Wood

Average waste generation in East Asia and Pacific's urban areas is substantially higher than national averages, at 0.96 kilogram per capita per day.

The majority of waste in the East Asia and Pacific region is organic (figure 3.2). Dry recyclables comprise one-third of waste. Many initiatives have emerged to recover usable materials from waste in the East Asia and Pacific region.

Waste Collection

At a national level, waste collection coverage in East Asia and Pacific averages about 71 percent (figure 3.3). Rates are highest in urban areas, at about 77 percent, and lower in rural communities at 45 percent. High-income countries and economies, such as Singapore; Hong Kong SAR, China; Japan; and Korea collect almost 100 percent of waste (figure 3.4). Where services exist in East Asia and Pacific, the majority of waste is collected on a door-to-door basis (in 18 out of 25 countries studied).

The informal sector is active in the region, with up to an estimated 200,000 active waste pickers in Beijing, China, and 16,000 in Ho Chi Minh City, Vietnam (Li 2015; CCAC, n.d.). Waste picker services are formalized in certain cities. For example, as part of the rehabilitation of the Baruni

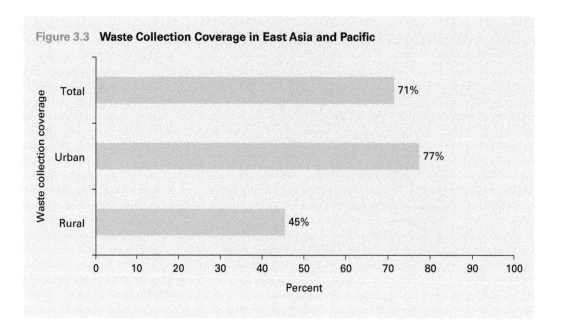

Figure 3.3 Waste Collection Coverage in East Asia and Pacific

Disposal Facility in the capital of Papua New Guinea, waste-picking activities have been regulated. In Port Vila, the capital of Vanuatu, waste pickers must register at the disposal facility to collect waste.

Within the East Asia and Pacific cities studied, source separation is commonly practiced (figure 3.5). The most commonly source-separated materials are paper and cardboard, cans and metals, plastics and packaging, and glass.

Waste Transportation

The distance traveled between city centers and final disposal sites ranges from 6 kilometers in Honiara, Solomon Islands, to 64 kilometers in Seoul, Korea. Waste transportation distances can be higher for cities with dense suburban populations and limited access to land outside urban centers.

Waste Disposal

In East Asia and Pacific, 46 percent of waste is disposed of in landfills (figure 3.6). Notably, slightly more than one-fifth of waste is incinerated in modern facilities. Incineration is typically practiced by high-income countries and economies with limited land availability, such as Japan (80 percent); Taiwan, China (64 percent); Singapore (37 percent); and Korea (25 percent), but has also become common practice in China (30 percent). Open dumping is relatively uncommon compared with other

Photo 3.1   **Landfill in China**

Photo 3.2   **A Waste and Street Cleaning Worker in Hoi An, Vietnam**

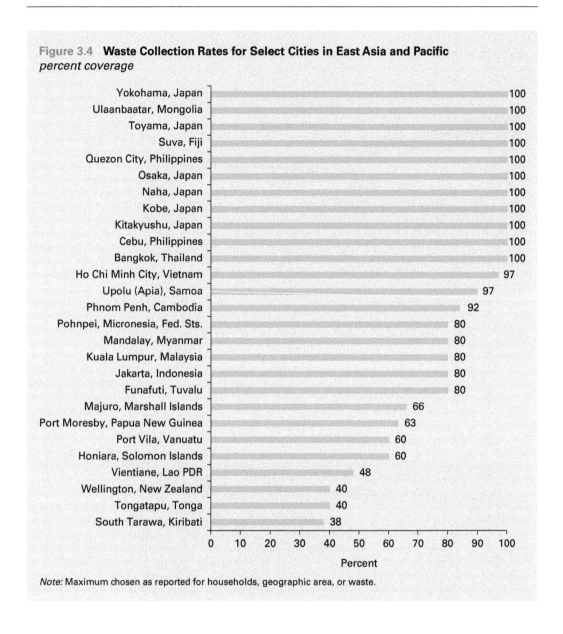

Figure 3.4 Waste Collection Rates for Select Cities in East Asia and Pacific
percent coverage

Note: Maximum chosen as reported for households, geographic area, or waste.

regions, potentially because of the advanced waste practices of highly populous, high-income economies within the region.

Many cities practice recycling to some extent. In East Asia and Pacific, 23 out of 28 cities with reported data recycle some amount of waste. Composting is developing as a practice in high-income or densely populated cities such as Wellington, New Zealand; Bangkok, Thailand; and Seoul, Korea.

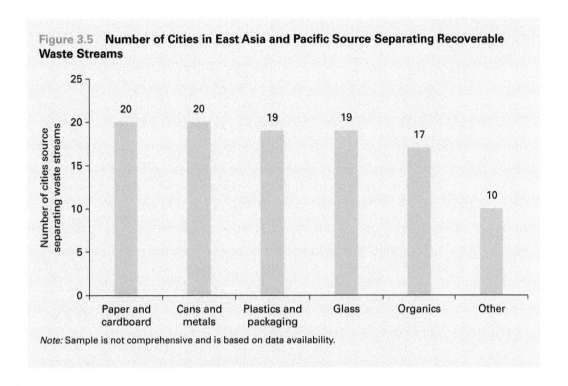

Figure 3.5   Number of Cities in East Asia and Pacific Source Separating Recoverable Waste Streams

Note: Sample is not comprehensive and is based on data availability.

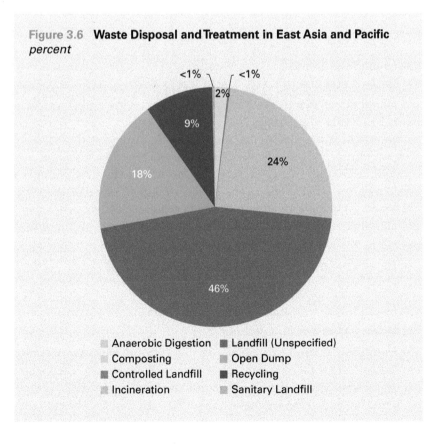

Figure 3.6   Waste Disposal and Treatment in East Asia and Pacific
percent

Europe and Central Asia

Key Insights

- Europe and Central Asia generated 392 million tonnes of waste in 2016, or 1.18 kilograms per capita per day.
- Waste collection rates in the region average 90 percent. The average urban collection rate is 96 percent, and the average rural rate is 55 percent.
- About three-quarters of waste in Europe and Central Asia has the potential to be recovered through recycling or organics management. Currently, 31 percent of waste materials are recovered through recycling and composting.
- Incineration is used to process 18 percent of waste across Europe and Central Asia, though the practice has mainly been adopted in Western European countries.
- Because of greater economic development and stringent European Union legislation, many of the higher recycling and collection rates in Europe and Central Asia occur in Western Europe.
- The focus in Europe and Central Asia is typically on improvement of waste collection systems, construction of central disposal sites, and closure of dumpsites.

Background and Trends

The Europe and Central Asia region includes 57 countries spanning Greenland in the west to the Russian Federation in the east. The region housed 912 million people in 2016. Waste prevention and recycling are increasing in the region. In European Union member states in Western Europe, targets for waste disposal and recycling are guided by legislation. To fulfill their membership requirements, new European Union member states are focused on increasing rates of waste diversion from landfills and are taking measures to achieve a 100 percent rate of sanitary waste disposal, if not achieved already. Much of the fastest growth in modernization of waste management systems is occurring in Eastern Europe and Central Asia, where governments are largely focused on closing old dumpsites and building centralized facilities for treatment and disposal.

Waste Generation and Composition

The Europe and Central Asia region generated 392 million tonnes of waste in 2016, or 1.18 kilograms per person each day (figure 3.7). The highest per capita waste generators are found in a few island states with high levels of tourism and in the economic hubs in Western Europe. The countries

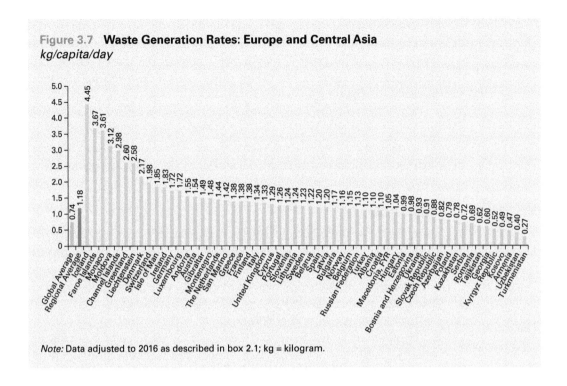

Figure 3.7   Waste Generation Rates: Europe and Central Asia
kg/capita/day

Note: Data adjusted to 2016 as described in box 2.1; kg = kilogram.

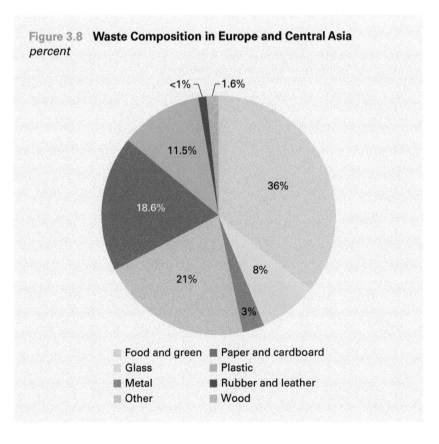

Figure 3.8   Waste Composition in Europe and Central Asia
percent

- Food and green
- Glass
- Metal
- Other
- Paper and cardboard
- Plastic
- Rubber and leather
- Wood

generating the least waste are largely in Eastern Europe or Central Asia and typically have a lower gross domestic product per capita. Urban areas generated 1.28 kilograms of waste per capita per day. However, in many countries in the region, per capita waste generation levels hardly differ between urban and rural areas.

Waste in the Europe and Central Asia region is mostly organic, as is consistent with global trends (figure 3.8). The region is only exceeded in its generation of solid recyclables, such as paper and plastic, by North America. In urban areas, waste composition is similar to national waste composition, with a slightly lower proportion of organic waste.

Waste Collection

Nationally, waste collection coverage is relatively high, at 90 percent (figure 3.9). The urban waste collection coverage rate of 96 percent exceeds the rural waste collection rate of 55 percent. In many European cities waste collection has been modernized with comprehensive truck fleets and planned systems (figure 3.10); in rural areas, however, waste collection systems are still in development.

In urban areas in Europe and Central Asia, waste collection typically takes place through a mix of door-to-door curbside collection and drop-offs at centralized bins.

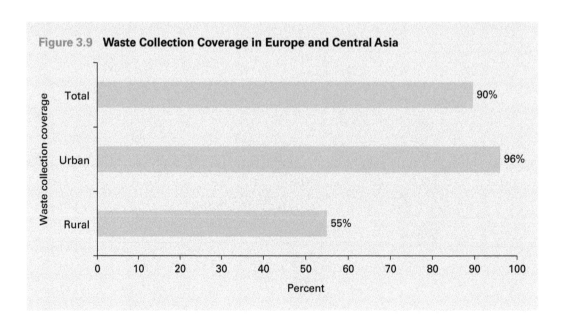

Figure 3.9 Waste Collection Coverage in Europe and Central Asia

Figure 3.10 Waste Collection Rates for Select Cities in Europe and Central Asia
percent coverage

Note: Maximum chosen as reported for households, geographic area, or waste.

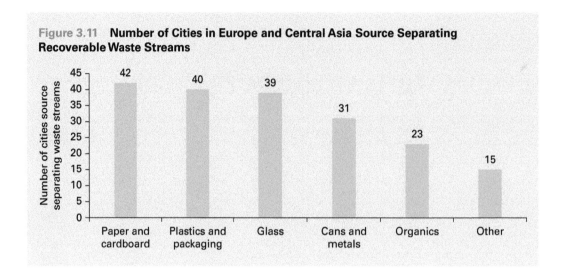

Figure 3.11 Number of Cities in Europe and Central Asia Source Separating Recoverable Waste Streams

Out of the 45 cities in the study that reported some type of source separation, the streams of waste that are most commonly source separated are paper and cardboard, plastics and packaging, and glass (figure 3.11).

Waste Transportation

In cities in Europe and Central Asia, the distance that waste is transported between main city centers and landfills ranges considerably, from 2 to 51 kilometers. Of 22 cities with reported data, 10 aggregate waste at a central transfer station or collection point before final disposal in landfills.

Waste Disposal

In Europe and Central Asia, one-quarter of waste is disposed of in some type of landfill (figure 3.12). Incineration, which accounts for 18 percent of waste disposal, is largely practiced in Western European countries with high technological capacity, advanced environmental regulations, and enforcement authority. Several countries have achieved high rates of recycling and composting (table 3.1).

Photo 3.3   **Recycling Plant in Bosnia and Herzegovina**

Figure 3.12   **Waste Disposal and Treatment in Europe and Central Asia**
percent

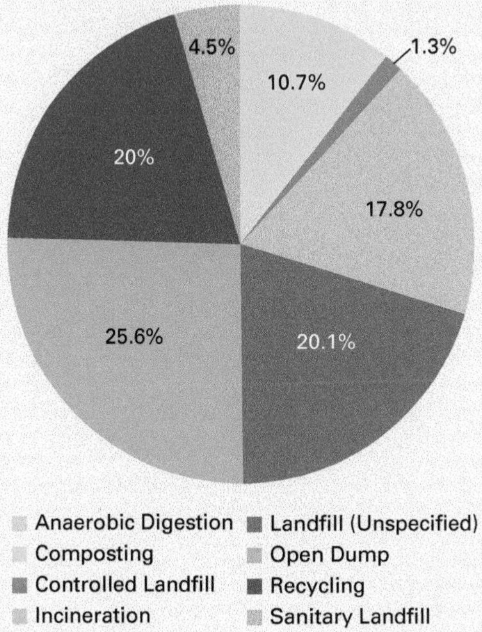

- 4.5%
- 10.7%
- 1.3%
- 17.8%
- 20%
- 25.6%
- 20.1%

Legend:
- Anaerobic Digestion
- Composting
- Controlled Landfill
- Incineration
- Landfill (Unspecified)
- Open Dump
- Recycling
- Sanitary Landfill

Table 3.1   Countries with High Recycling and Composting Rates in Europe and Central Asia

	Recycling rate (percent)		Composting rate (percent)
Faroe Islands	67	Austria	31
Liechtenstein	64	Netherlands	27
Iceland	56	Liechtenstein	23
Isle of Man	50	Switzerland	21
Germany	48	Luxembourg	20
Slovenia	46	Belgium	19
San Marino	45	Denmark	19
Belgium	34	Germany	18
Ireland	33	Italy	18
Sweden	32	France	17

Note: Rates represent percentage of total waste.

Latin America and the Caribbean

Key Insights

- The Latin American and the Caribbean region generated 231 million tonnes of waste in 2016, with an average of 0.99 kilogram per capita per day.
- Some 52 percent of municipal solid waste is classified as food and green waste.
- Waste collection coverage for the region is relatively comprehensive at 84 percent, on average, although average coverage for rural areas is 30 percent.
- About 69 percent of waste is disposed of in some form of landfill, and more than 50 percent of waste is disposed of in sanitary landfills with environmental controls.
- The region recycles 4.5 percent of waste.
- Some countries are pursuing opportunities to recover energy from waste through methods such as landfill gas collection and anaerobic digestion.

Background and Trends

The Latin America and the Caribbean region consists of 42 countries that include South America and the Caribbean Islands. The region had a population of 638 million in 2016. Solid waste systems in the region are in the process of modernization, though practices vary based on income level. At an urban level, many cities have initiated source-separation programs, and recycling rates are highest for materials such as aluminum, paper, and plastic. Recycling is common in the region except in the Caribbean islands, where recycling markets are nascent.

An increasing amount of waste is being disposed of in sanitary landfills, with or without environmental and social controls, but a significant amount of waste is still dumped, burned, or used as animal feed. The stability of dumpsites is an issue, especially given the frequency of natural disasters in the Caribbean. The Caribbean is also more prone to plastic marine litter washing up onshore and needs to address that challenge as well. Some advanced cities are beginning to convert landfill gas to energy. Other cities are exploring new technologies such as waste-to-energy incineration and anaerobic digestion, with anaerobic digestion receiving particular attention. From a policy perspective, most countries and cities have at least one regulatory mechanism in place to guide waste management activities.

Cost recovery varies across the region and includes government subsidies, taxes, user fees, and cross-subsidization across income levels. Environmental levies for littering are common, but it is not always clear that the funds are used for solid waste management activities.

Waste Generation and Composition

The Latin America and the Caribbean region generated 231 million tonnes of waste in 2016, at an average of 0.99 kilogram per person each day (figure 3.13). Many of the highest waste generators are island states with active tourist economies.

Figure 3.13   Waste Generation Rates: Latin America and the Caribbean Region
kg/capita/day

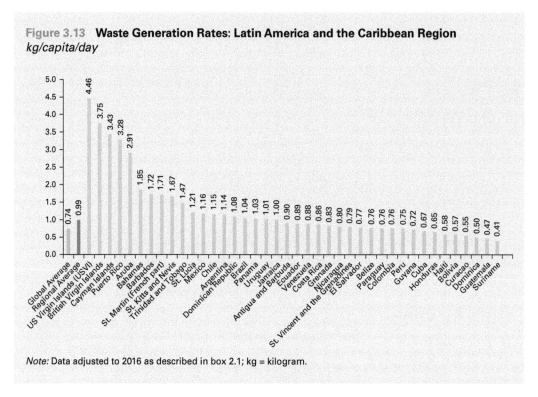

Note: Data adjusted to 2016 as described in box 2.1; kg = kilogram.

Figure 3.14   Waste Composition in Latin America and the Caribbean
percent

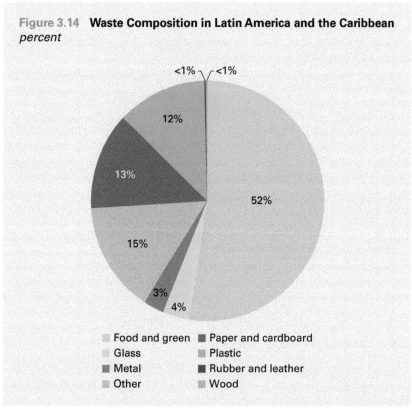

About half of the waste in the Latin America and the Caribbean region is food and green waste (figure 3.14). About one-third of waste is composed of dry recyclables. It is likely that the almost 15 percent of waste that has not been characterized through formal systems is largely organic, given that areas outside the purview of municipal waste systems tend to be rural or lower in income, and these areas tend to generate more wet or organic waste.

Waste Collection

Waste collection coverage is quite high for the Latin America and the Caribbean region compared with global trends. At an urban level, about 85 percent of waste is collected (figure 3.15), and most waste collection systems in Latin America and the Caribbean are on a door-to-door basis. In rural communities, waste collection coverage is about 30 percent. Collection coverage varies significantly, with coverage of greater than 95 percent in cities in countries such as Uruguay and Colombia, and as low as 12 percent in Port au Prince, Haiti (figure 3.16). The informal sector is highly active within the region. Cities studied reported varying numbers of active waste pickers, ranging from 175 in Cusco, Peru, to 20,000 in São Paolo, Brazil (Lizana 2012; CCAC, n.d.). Some large Latin American cities average almost 4,000 active waste pickers collecting recyclable materials.

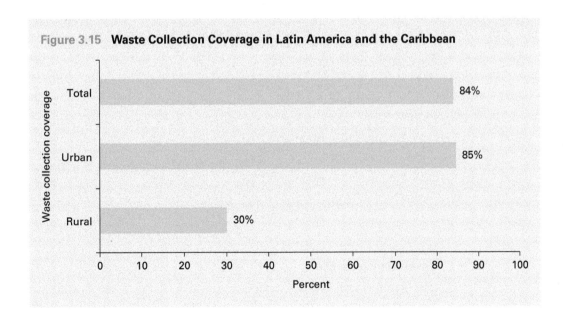

Figure 3.15   Waste Collection Coverage in Latin America and the Caribbean

Figure 3.16 Waste Collection Rates for Select Cities in Latin America and the Caribbean
percent coverage

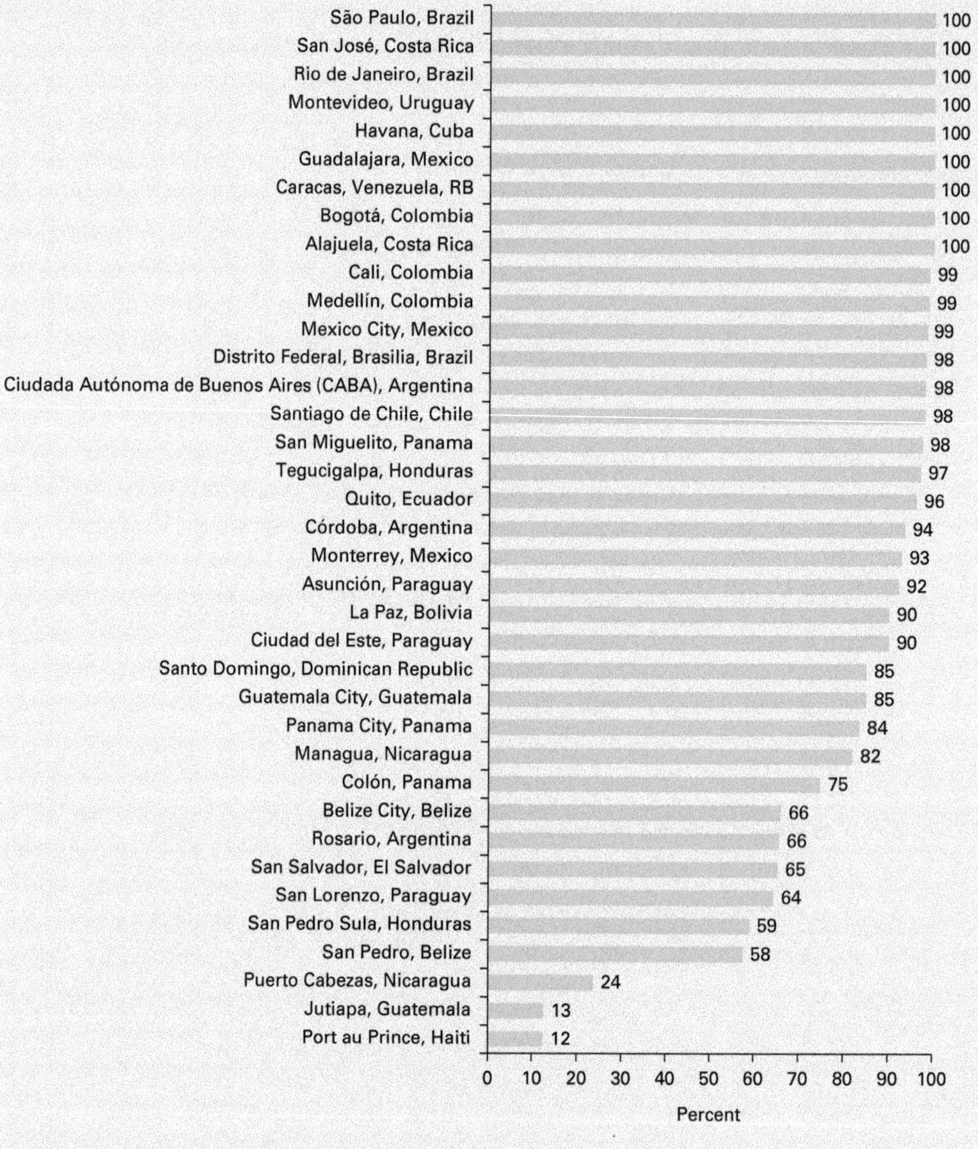

City	Percent
São Paulo, Brazil	100
San José, Costa Rica	100
Rio de Janeiro, Brazil	100
Montevideo, Uruguay	100
Havana, Cuba	100
Guadalajara, Mexico	100
Caracas, Venezuela, RB	100
Bogotá, Colombia	100
Alajuela, Costa Rica	100
Cali, Colombia	99
Medellín, Colombia	99
Mexico City, Mexico	99
Distrito Federal, Brasilia, Brazil	98
Ciudada Autónoma de Buenos Aires (CABA), Argentina	98
Santiago de Chile, Chile	98
San Miguelito, Panama	98
Tegucigalpa, Honduras	97
Quito, Ecuador	96
Córdoba, Argentina	94
Monterrey, Mexico	93
Asunción, Paraguay	92
La Paz, Bolivia	90
Ciudad del Este, Paraguay	90
Santo Domingo, Dominican Republic	85
Guatemala City, Guatemala	85
Panama City, Panama	84
Managua, Nicaragua	82
Colón, Panama	75
Belize City, Belize	66
Rosario, Argentina	66
San Salvador, El Salvador	65
San Lorenzo, Paraguay	64
San Pedro Sula, Honduras	59
San Pedro, Belize	58
Puerto Cabezas, Nicaragua	24
Jutiapa, Guatemala	13
Port au Prince, Haiti	12

Percent

Note: Maximum chosen as reported for households, geographic area, or waste.

Waste Transportation

The main transportation mode for waste in Latin America and the Caribbean is trucks. Of the 21 cities reporting waste transportation practices, 16 aggregate waste at a transfer station or other site before final disposal, and most cities have transfer stations in operation, ranging from

1 in Cusco, Peru, to 12 in Mexico City, Mexico. Aggregation stations may be intended either to transfer waste to larger trucks or to increase efficiency of materials recovery. These stations may be formal or informal recycling centers where waste pickers sort materials for recycling. After waste is collected, the distance traveled between city centers and final disposal sites ranges from 4 to 62 kilometers.

Waste Disposal

More than two-thirds of waste in the Latin America and the Caribbean region is disposed of in some type of landfill (figure 3.17), although some of these may be well-run dumps. More than half of waste is disposed of in sanitary landfills with some environmental controls, reflecting a general regional focus on sustainable disposal methods. Open dumping accounts for about 27 percent of waste disposal and treatment. Recycling and composting systems are emerging across the region, although the extent of implementation varies by country. Many cities are focused on recovering waste; for example, cities such as Montevideo, Uruguay, and Bogotá and Medellín, Colombia, recycle more than 15 percent of waste. In addition, cities such as Mexico City, Mexico, and Rosario, Argentina, compost more than 10 percent of waste. Landfill gas collection has emerged as the main mechanism of recovering energy from waste in Latin America and the Caribbean.

Photo 3.4   **Plastic Bottle Collection in Jamaica**

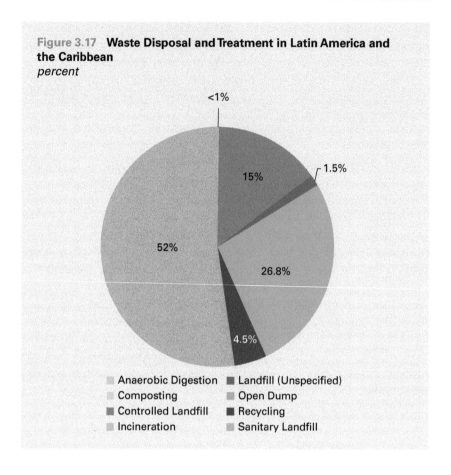

Figure 3.17   Waste Disposal and Treatment in Latin America and the Caribbean
percent

Legend:
- Anaerobic Digestion
- Composting
- Controlled Landfill
- Incineration
- Landfill (Unspecified)
- Open Dump
- Recycling
- Sanitary Landfill

Middle East and North Africa

Key Insights

- The Middle East and North Africa region generated 129 million tonnes of waste in 2016, the lowest total of any region, primarily because of its lower population, at an average of 0.81 kilogram per capita daily. However, the region will double waste generation by 2050.
- Some 53 percent of all waste is disposed of in open dumps although countries are seeking alternative methods to dispose of waste, especially in the Gulf Cooperation Council (GCC). Recycling and composting are widespread at a pilot scale.
- In GCC countries, waste collection rates are nearly 100 percent, and these countries are exploring ways to recover value from waste through waste-to-energy projects.
- Political instability in certain countries has hindered the development of formal waste systems in many areas. However, citizen engagement initiatives are strong and governments are increasingly pursuing reforms, integration of the private sector, and improved fee recovery.
- Legal and institutional reform has become a common focus for many nations, such as Jordan, Morocco, and the GCC countries.

Background and Trends

The Middle East and North Africa region consists of 21 countries spanning Morocco in the west to the Islamic Republic of Iran in the east. The region was home to a population of 437 million people in 2016. Waste generation and management practices in the region vary widely. Countries such as Bahrain, the United Arab Emirates, and Kuwait generate more than 1.5 kilograms of waste per person per day, while countries such as Morocco, Djibouti, and the Republic of Yemen generate less than 0.6 kilogram per person per day.

Although political fragility has affected the delivery of services and progress of the solid waste sector in several areas, many governments are making efforts to address waste challenges through legal, technical, and institutional reforms. Strong citizen engagement initiatives and financial investments are underway, such as in Morocco (box 3.1), and several countries are working to integrate the private sector into service delivery and to increase fee recovery. Many high-income countries in the GCC are finding sustainable ways to dispose of waste, including through waste-to-energy projects, and several waste infrastructure projects are being tendered or are under construction in the region. Most GCC countries are also developing new regulations and institutional structures.

Waste Generation and Composition

Waste generation in the Middle East and North Africa region is relatively modest compared with global trends, primarily due to a lower population. The region generated 129 million tonnes of waste in 2016, at an average of 0.81 kilogram per person per day (figure 3.18). However, the waste generation rate in cities is significantly higher, at an average of 1.38 kilograms per person per day. Many of the largest waste generators are high-income countries, mainly those in the GCC.

Figure 3.18   Waste Generation Rates: Middle East and North Africa Region
kg/capita/day

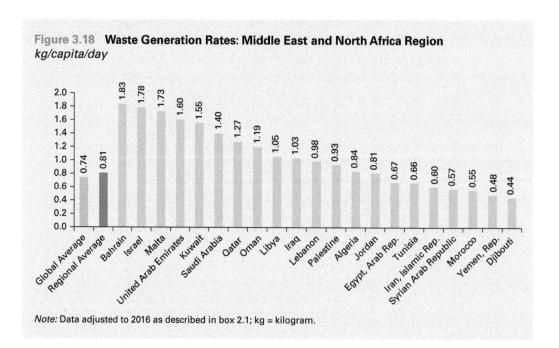

Note: Data adjusted to 2016 as described in box 2.1; kg = kilogram.

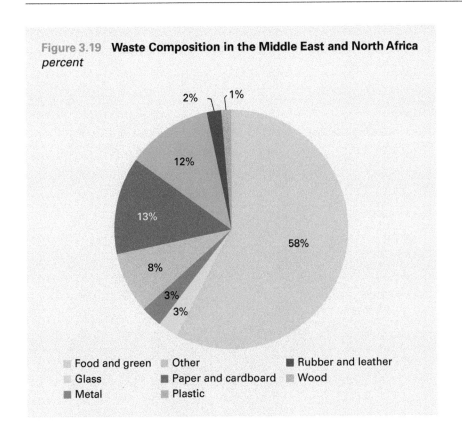

Figure 3.19 Waste Composition in the Middle East and North Africa
percent

Food and green	Other	Rubber and leather
Glass	Paper and cardboard	Wood
Metal	Plastic	

Food and green waste, at 58 percent, is the predominant type of waste in the Middle East and North Africa (figure 3.19). About one-third of waste is composed of dry recyclables, and the rising recycling activity in the region reflects an increasing willingness to capture the value of these materials. A major focus in the region is on reducing food waste and addressing organic waste management.

Waste Collection

Waste collection coverage is relatively comprehensive for the Middle East and North Africa region. Coverage is highest in urban areas, with an average of 90 percent of waste being collected (figure 3.20). Many cities reported 100 percent collection coverage (figure 3.21). Rural coverage is relatively high for most countries, with an average of 74 percent of waste being collected, though there is significant variation between countries. In Qatar, 100 percent of rural waste is collected. However, in Tunisia and the Arab Republic of Egypt, 5 percent and 15 percent of rural waste is collected, respectively. The informal sector is active across the Middle East and North Africa region. For example, an estimated 96,000 informal waste pickers are active in Cairo and account for 10 percent of the waste collected in the city (IFC 2014).

Photo 3.5   One Form of Waste Collection in West Bank

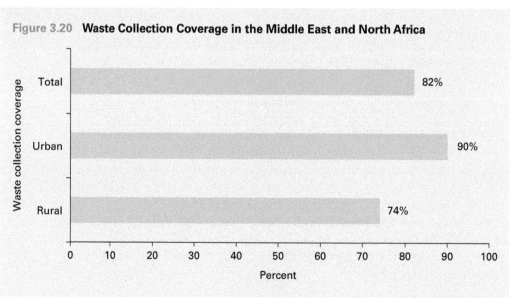

Figure 3.20   Waste Collection Coverage in the Middle East and North Africa

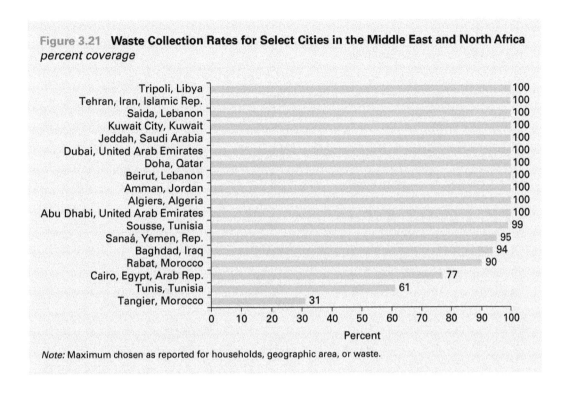

Figure 3.21 **Waste Collection Rates for Select Cities in the Middle East and North Africa**
percent coverage

Note: Maximum chosen as reported for households, geographic area, or waste.

Where waste collection exists in urban areas, the predominant method is door-to-door pickup by trucks. Source separation is not common within the region.

Waste Transportation

The distance traveled between city centers and final disposal sites ranges significantly, from 3 to 40 kilometers (table 3.2). It is common to aggregate waste at a transfer station or another site before final disposal, and many cities have transfer stations in operation. However, many cities studied stated that the number of transfer stations in operation fell short of the optimal number for the urban waste management system.

Waste Disposal

Waste disposal practices vary widely in the Middle East and North Africa region. Open dumping is prevalent, at 53 percent of total waste management (figure 3.22). For example, about 940 dumps exist in Lebanon for municipal waste, as well as construction and demolition waste (Republic of Lebanon Ministry of Environment and UNDP 2017). Furthermore, in high-income countries, most landfills are not engineered landfills and effectively operate as dumps.

Table 3.2 Examples of Transfer Station Availability and Transportation Distance in the Middle East and North Africa

City	Number of transfer stations in operation	Number of transfer stations needed	Distance from city center to final disposal site (km)	Population (1000s)
Tangier, Morocco	0	2	20	1,100
Sanaá, Yemen, Rep.	1	5	17	2,331
Rabat, Morocco	1	No data	20	650
Beirut, Lebanon	2	2	9	650
Sfax, Tunisia	2	4	40	300
Amman, Jordan	3	No data	24	2,400
Abu Dhabi, United Arab Emirates	6	No data	No data	1,145
Tunis, Tunisia	8	10	15	700
Baghdad, Iraq	9	No data	No data	7,000
Tehran, Iran, Islamic Rep.	11	11	39	8,432
Ramallah, West Bank and Gaza	No data	2	No data	
Damascus, Syrian Arab Republic	No data	11	No data	2,566
Saida, Lebanon	No data	No data	3	150
Beni Mellal, Morocco	No data	No data	4.5	192
Jeddah, Saudi Arabia	No data	No data	40	4,076

However, landfill usage is rapidly increasing. For example, the percentage of municipal solid waste collected and disposed of in sanitary landfills in Morocco has increased from 10 percent in 2008 to 32 percent in 2012 and to 53 percent in 2016, and once five new sanitary landfills that are currently under construction are complete, the rate is expected to reach 80 percent (Sarraf 2018, personal communication, May 30, 2018). Many countries in the GCC, such as Qatar, have integrated higher recycling and waste-to-energy rates into national plans and have begun construction (Qatar General Secretariat for Development and Planning 2011). Several high-income GCC countries are pursuing waste-to-energy solutions and are planning properly designed waste management facilities, including incinerators and sanitary landfills.

Recycling and composting systems are increasing in prevalence. Of the 21 countries studied, 16 have reported some amount of recycling activities. The share of waste recovered is typically low, about 1–8 percent, but ranges up to 25 percent. Out of 21 countries, 9 have reported some level of composting activity.

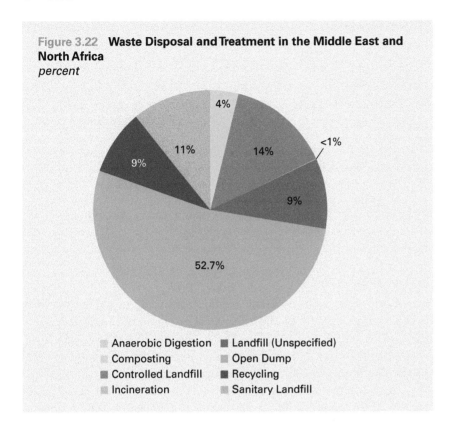

Figure 3.22 **Waste Disposal and Treatment in the Middle East and North Africa**
percent

Legend:
- Anaerobic Digestion
- Composting
- Controlled Landfill
- Incineration
- Landfill (Unspecified)
- Open Dump
- Recycling
- Sanitary Landfill

North America

Key Insights

- The North American region generates the highest average amount of waste, at 2.21 kilograms per capita per day; total waste generated was 289 million tonnes annually in 2016.
- Waste collection coverage in North America is nearly universal, at 99.7 percent, with the gap in collection coverage occurring in Bermuda.
- More than 55 percent of waste is composed of recyclables including paper, cardboard, plastic, metal, and glass.
- At 54 percent, more than half of waste in North America is disposed of at sanitary landfills and one-third of waste is recycled.

Background and Trends

North America is the smallest region, consisting of three countries: Bermuda, Canada, and the United States. The region was home to a population of 359 million in 2016. The United States is the largest of the three, with a population of 322 million, and Bermuda is the smallest with 62,000 people. All three countries in North America are high-income nations, and as such, waste management and disposal practices tend to be advanced relative to global trends. Waste management systems generally operate in an environmentally sound manner, have high capacity, serve nearly all citizens, and enjoy more consistent financial stability and fee collection than systems in lower-income countries.

Given more advanced information management systems, data availability and accuracy are relatively strong in North America.

Waste Generation and Composition

Though home to less than 5 percent of the global population in 2016, North America generated 14 percent of the world's waste, at 289 million tonnes with a daily rate of 2. 21 kilograms per capita (figure 3.23). The high waste generation rate reflects the high-income status of these countries and related economic activity. Waste generation rates in North American cities vary, with cities such as Seattle in the United States generating up to 3.13 kilograms per capita per day and others such as Ottawa, Canada, generating only 0.95 kilogram per capita per day (Seattle Public Utilities 2016; City of Ottawa 2018).

As an island state with high tourist activity, Bermuda is the highest waste generator per capita in North America. Canada generates the least amount of waste, though it is not far behind the United States on a per capita basis.

The composition of waste in North America is diverse. Unlike other regions, food and green waste accounts for less than 30 percent of the total waste stream (figure 3.24). More than 55 percent of waste is dry recyclables. Paper and cardboard comprise 28 percent of total waste, and plastic 12 percent.

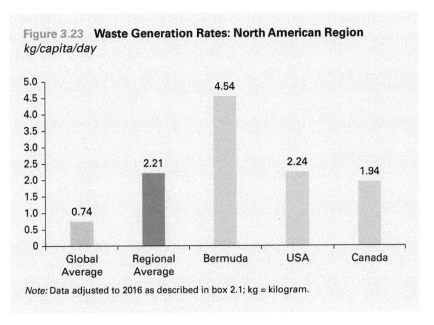

Figure 3.23 **Waste Generation Rates: North American Region**
kg/capita/day

Note: Data adjusted to 2016 as described in box 2.1; kg = kilogram.

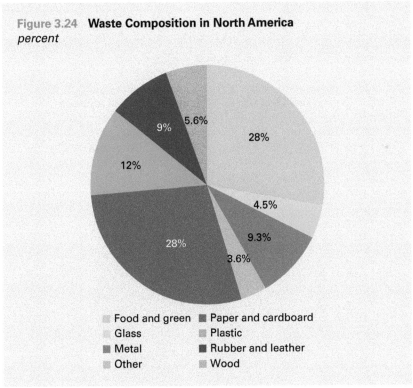

Figure 3.24 **Waste Composition in North America**
percent

Food and green — Paper and cardboard
Glass — Plastic
Metal — Rubber and leather
Other — Wood

Waste Collection

Some 99.7 percent of waste in North America is collected, with a gap only in Bermuda. Waste is typically collected at the curb beside residential or commercial establishments. Rural areas have access to formal collection systems, which are typically either door-to-door services or central drop-off points managed by municipalities or private entities. Local laws in many cities mandate that waste collection be provided to households, or that larger residential buildings or commercial institutions contract with private operators for waste collection. Collection is typically done with trucks and, depending on the location of the treatment or disposal facility, waste is often shifted to larger trucks at transfer stations or sorting facilities to more effectively transport the large quantities of waste over long distances.

Waste Disposal

Over half of waste in North America is disposed of in sanitary landfills (figure 3.25). One-third of waste is recycled, about 12 percent is incinerated in incinerators with energy recovery, and less than 1 percent is composted. In North America, landfills are highly regulated and waste is disposed of in an environmentally sound manner. U.S. landfills are regulated by the national Environmental Protection Agency, and municipal solid waste landfills must be designed with environmental controls and report on certain performance measures, such as methane generation, as well as the quality of the air, water, and soil.

As an island nation, Bermuda faces unique land constraints. In 1994, a central waste-to-energy incinerator was constructed. The facility reduces waste volume by 90 percent and currently 67 percent of the country's waste is incinerated (Government of Bermuda n.d.).

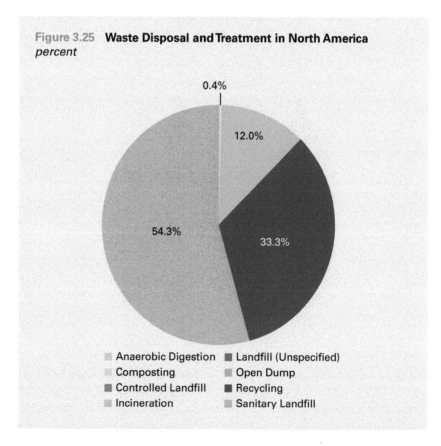

Figure 3.25 Waste Disposal and Treatment in North America
percent

- 0.4%
- 12.0%
- 33.3%
- 54.3%

Legend:
- Anaerobic Digestion
- Composting
- Controlled Landfill
- Incineration
- Landfill (Unspecified)
- Open Dump
- Recycling
- Sanitary Landfill

South Asia

Key Insights

- The South Asia region generated 334 million tonnes of waste in 2016, at an average of 0.52 kilogram per capita daily, including both urban and rural waste. Total waste generation is expected to double in the region by 2050.
- About 57 percent of waste in South Asia is characterized as food and green waste.
- About 44 percent of waste is collected in South Asia, mainly through door-to-door systems.
- About three-fourths of waste is currently openly dumped in South Asia, although improvements to collection systems and construction of sanitary final disposal sites are underway.

Background and Trends

The South Asia region has only eight countries but a large population. The three population hubs of India, Pakistan, and Bangladesh together have a population of 1.68 billion people; Afghanistan, Nepal, and Sri Lanka are home to nearly 85 million people; and the smaller states of Bhutan and Maldives have about 1.2 million people. The countries in South Asia are diverse not only in population, but also in economic development and geography.

Almost all cities in the South Asia region practice some open dumping, but cities are increasingly developing sanitary landfills and pursuing recycling. Most cities hire private contractors or nongovernmental organizations to collect waste from neighborhoods and institutions and pay collectors based on the amount of waste transported to disposal sites. Although rules and regulations have been developed at national and state levels, these criteria are still being translated into practice and accountability structures at the city level.

Implementation of policies is challenging because of a lack of enforcement mechanisms. In addition to improving legal enforcement, strengthening the technical and institutional capacity of administrators at all levels of solid waste management systems, from municipal staff to the regulators and operators, is a common priority.

Waste Generation and Composition

The South Asia region generated 334 million tonnes of waste in 2016, at an average of 0.52 kilogram per person each day (figure 3.26). Rural waste generation is significantly lower than urban waste generation and reduces the average amount generated in the region. The islands of

Maldives generate the most amount of waste per capita. In cities in South Asia, waste generation rates vary widely, with cities such as Kabul, Afghanistan, generating about 1.5 kilograms per capita per day, and cities such as Butwal, Nepal, generating only about 0.2 kilogram per capita per day (Asian Development Bank 2013; World Bank 2016).

Most waste in the South Asia region is organic (figure 3.27). A large proportion of waste is not classified, though it is assumed that most of this waste is inert. Waste cleaned from drains and silt is often mixed into the solid waste disposed of by municipalities. Construction and demolition waste is also often included in the data reported for South Asia though it will gradually be managed separately as a result of new rules in India established in 2016 (Ministry of Environment, Forest and Climate Change 2016).

Waste Collection

Excluding Maldives, Sri Lanka, and Afghanistan, where data were not available, urban waste collection coverage in South Asia is about 77 percent (figure 3.28), although coverage varies considerably by country and city (figure 3.29). Rural areas have lower collection coverage rates of about 40 percent.

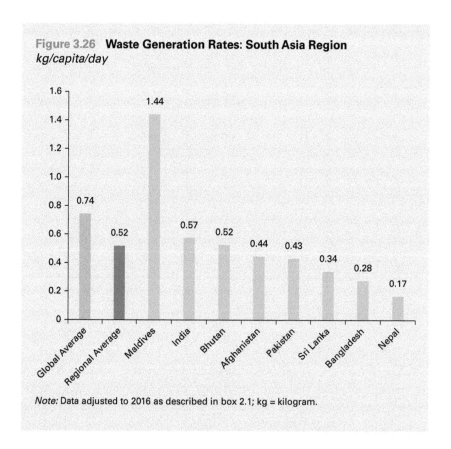

Figure 3.26 Waste Generation Rates: South Asia Region
kg/capita/day

Note: Data adjusted to 2016 as described in box 2.1; kg = kilogram.

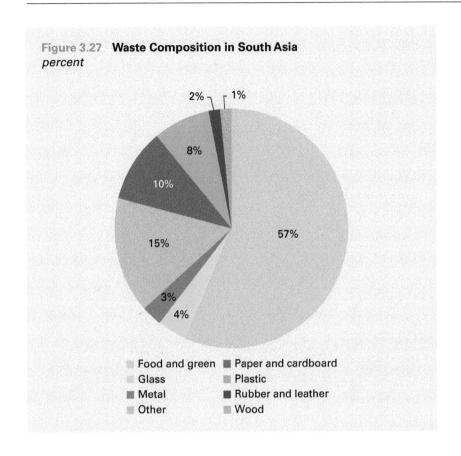

Figure 3.27   **Waste Composition in South Asia**
percent

- Food and green
- Glass
- Metal
- Other
- Paper and cardboard
- Plastic
- Rubber and leather
- Wood

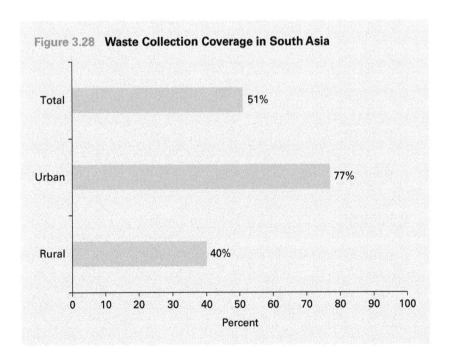

Figure 3.28   **Waste Collection Coverage in South Asia**

Figure 3.29   Waste Collection Rates for Select Cities in South Asia
percent coverage

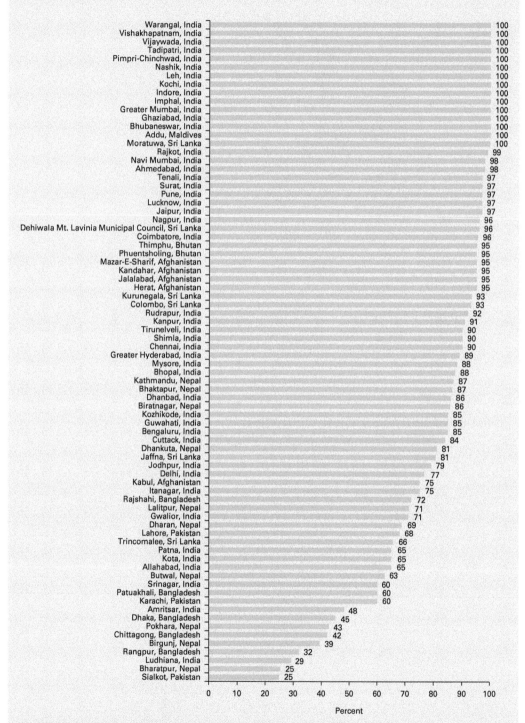

Note: Maximum chosen as reported for households, geographic area, or waste.

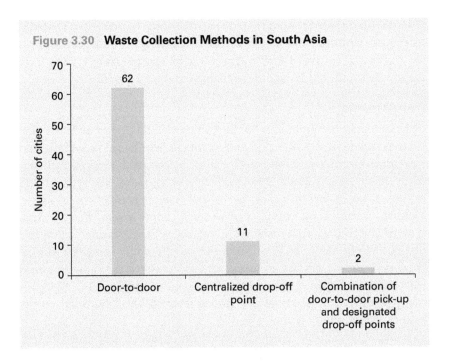

Figure 3.30   **Waste Collection Methods in South Asia**

Waste collection services, where they exist in cities, typically occur door-to-door (figure 3.30). In certain cities, such as Butwal, Nepal, and Kota, India, residents dispose of waste at a primary collection point, from which aggregated waste is transported to the final disposal site. This practice is extremely common, and designated primary collection sites or open plots of land often eventually become unofficial sites for dumping. In Navi Mumbai, India, a waste collector notifies residents to bring waste to the collection vehicle (India, Ministry of Urban Development 2016).

Informal waste collection and materials recovery activities are prolific in South Asia. Most cities studied reported between 150 and 1,100 active waste pickers. The large cities of Dhaka, Bangladesh, and Delhi, India, reported 120,000 and 90,000 active waste pickers, respectively. Unorganized waste pickers are commonly seen working at either informal or formal transfer stations. At landfills, waste pickers are typically organized or are part of a cooperative (Enayetullah and Hashmi 2006).

Waste Transportation

On average, waste is transported about 15.5 kilometers between city centers and final disposal sites in the South Asia region. Primary collection occurs in various ways, and the main forms of transportation are three-wheeled push carts, tractors, and bicycle rickshaws. Out of 53 cities, 38 reported aggregating waste at transfer stations or secondary collection points before transporting it to final sites. Transfer stations may be designated sites with minimal infrastructure or constructed with technologies for automated sorting. Many aggregation centers are simply temporary storage sites and mostly facilitate manual handling of waste.

Waste Disposal

Open dumping is common in South Asia (figure 3.31), and most existing landfills lack leachate collection and treatment, landfill gas collection, and sometimes even liners. However, the remediation of dumpsites and construction of formal landfills are actively taking place, and official and well-functioning facilities tend to be privately operated. For example, Maldives is mitigating dumping of waste by improving waste collection systems and constructing sustainable disposal sites that can serve multiple islands (World Bank 2017a). Four out of the eight countries recycle between 1 and 13 percent of waste, and seven out of the eight countries have begun composting programs to manage organic waste. Waste-to-energy incineration potential has gained interest, but substantial results have not yet been proven.

Initiatives to improve waste disposal began in India in 2014, and interest in other South Asian countries is growing. Many cities are establishing central authorities to increase capacity to plan and operate the waste management sector. The focus is on developing waste disposal strategies that include locally tailored and cohesive approaches. Depending on the locality, cities are navigating varied constraints related to land, capacity, availability of local operators, financing, and alignment of waste technology and waste composition, and more than one solution is needed. Waste management is

Photo 3.6 **Dumpsite in Sri Lanka**

increasingly recognized as not only a social, health, and environmental issue, but an economic one, in which waste recovered and land used wisely can generate financial savings. Indian cities can access funds, mainly from the Swachh Bharat Mission (box 3.2), to improve waste management programs.

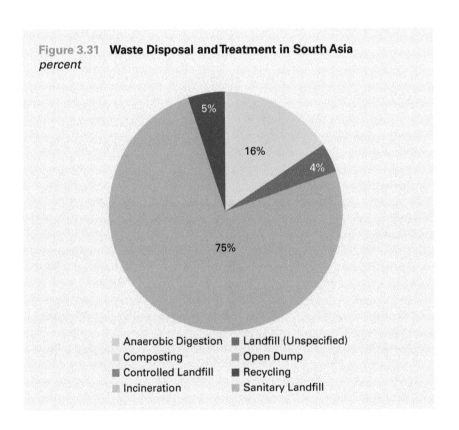

Figure 3.31 Waste Disposal and Treatment in South Asia
percent

- Anaerobic Digestion
- Composting
- Controlled Landfill
- Incineration
- Landfill (Unspecified)
- Open Dump
- Recycling
- Sanitary Landfill

Box 3.2   Swachh Bharat Mission (Clean India Mission)

The Swachh Bharat Mission is a national initiative in India to clean up cities, towns, and rural areas. Interventions range from cleaning roads and infrastructure to improving solid waste management and household sanitation practices. The government of India and involved stakeholders have supported actions in more than 4,000 cities, towns, and rural areas to date. Swachh Bharat is anticipated to fund more than $9.5 billion in investments and is providing incentives for jurisdictions to compete by publicly monitoring performance across cities.

Source: Swachh Bharat website (http://www.swachhbharaturban.in).

Sub-Saharan Africa

Key Insights

- The Sub-Saharan Africa region generated 174 million tonnes of waste in 2016, at a rate of 0.46 kilogram per capita per day. It is the fastest growing region, with waste expected to nearly triple by 2050.
- Waste in Sub-Saharan Africa is primarily organic, with 40 percent of it being food and green waste.
- Overall waste collection rates are about 44 percent, although the rate is much higher in urban areas than in rural areas, where waste collection services are minimal.
- About 69 percent of waste is openly dumped, although use of landfills and recycling systems is becoming more prevalent.
- The region is experiencing substantial growth and modernization, with a large focus on building sustainable final disposal sites, improving collection coverage, closing dumpsites, and providing environmental education for the public.
- Institutional setups for operations and maintenance and the regulatory framework are generally not clearly defined. National governments are increasingly delivering traditional municipal waste management services.

Background and Trends

The Sub-Saharan Africa region consists of 48 countries and was home to 1.03 billion people in 2016. It is one of the fastest growing regions of the world; more than half of the world's population growth is anticipated to occur in Sub-Saharan Africa by 2050 (United Nations 2017). With waste generation expected to more than quadruple by 2050, it is alarming that waste is predominantly openly dumped. Waste management systems are slowly improving as cities are more frequently prioritizing the construction of landfills, closure of dumps, and formalization of collection systems to improve the environmental and health impacts of waste.

Countries and cities are also increasingly focusing on solid waste management and environmental education to reduce waste generation and improve sorting and collection systems. Efforts are also being made to improve recycling systems to divert waste from dumps and final disposal sites and to increase employment for informal waste collection workers.

Population growth in Sub-Saharan Africa, amid high poverty rates, makes waste service fee collection and financing of the overall system key challenges for the region. Because governments have limited resources, waste often becomes a lower priority sector. However, governments are taking action to improve the financing of solid waste systems and are increasingly trying to find innovative tools and resources to address this issue. For example, in Senegal fiscal reform is aiming to designate funds for solid waste services.

Experience has shown that a lack of clarity institutionally or politically can impede the role of local governments in delivering solid waste management services and hinder partnerships with the private sector. National governments are now more commonly undertaking traditional municipal roles in delivery of solid waste management services. This pattern goes against the general global trend of decentralization and has led to mixed results.

Data collection systems for solid waste management are nascent in Sub-Saharan Africa and face significant gaps. However, data are increasingly being collected by municipal agencies, nonprofits, private operators, consulting firms, and other local organizations and cooperatives. Though not all data are published online, public data availability is anticipated to improve substantially in the near future.

Waste Generation and Composition

The Sub-Saharan Africa region generated 174 million tonnes of waste in 2016, or 0.46 kilogram per person each day (figure 3.32). The region's per capita generation rate is the lowest globally. The largest waste generators are typically middle-income countries or those with significant tourist populations. Waste generation in urban areas, at 0.74 kilogram per person each day, is higher than the regional average, a situation that could be linked to higher income and tourism activity.

More than 40 percent of the waste in the Sub-Saharan Africa region is organic (figure 3.33) and 30 percent of waste is typically inert waste such as sand and fine particles. Typical consumption patterns in the region are changing and moving toward more packaged products and electronics. An increase in imports is also leading to larger amounts of packaging.

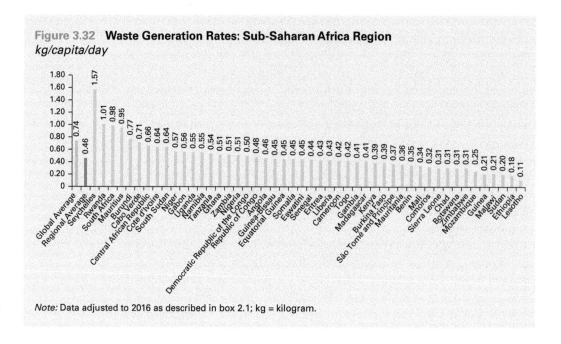

Figure 3.32   Waste Generation Rates: Sub-Saharan Africa Region
kg/capita/day

Note: Data adjusted to 2016 as described in box 2.1; kg = kilogram.

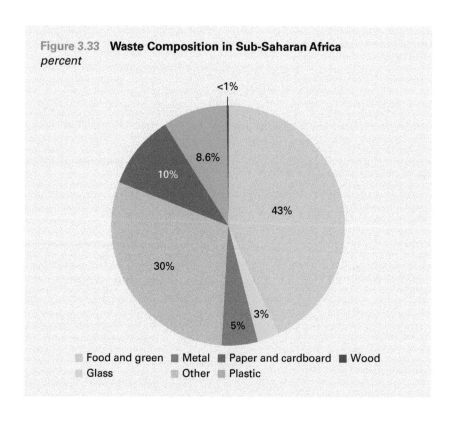

Figure 3.33   Waste Composition in Sub-Saharan Africa
percent

Waste Collection

Less than half of the waste generated in Africa is collected formally (figure 3.34). Collection coverage is much more comprehensive in urban areas than in rural areas, where collection is often nonexistent. Because of moderate formal collection rates, open dumping and burning are commonly pursued to eliminate remaining household waste. Often, waste that is formally collected is still disposed of at central dumpsites. The informal sector is active in many African cities and is largely responsible for recycling.

Waste collection systems are typically more developed in cities, and many cities in Sub-Saharan Africa have collection rates of more than 50 percent (figure 3.35).

Many cities use a dual system in which waste is first collected door to door and later from a centralized point at which collected waste is aggregated. Among cities with available data, 88 percent reported that waste collection typically occurs on a door-to-door basis. Other neighborhoods, including unplanned neighborhoods, have designated areas or dumpsters in which residents can deposit their waste. However, littering is a major issue in most cities. Despite having bins or dumpsters, it is common to see waste disposed of haphazardly. Almost every country in the region is at a very early stage in source separation. Efforts are often led by the private sector and nongovernmental organizations in the main capital cities at both household and commercial levels to improve the purity of waste streams and cost recovery.

It is typical to see higher waste collection rates in formal settlements and high-income areas, where it is easier to collect waste as a result of road development and greater density of housing. Most waste collection systems

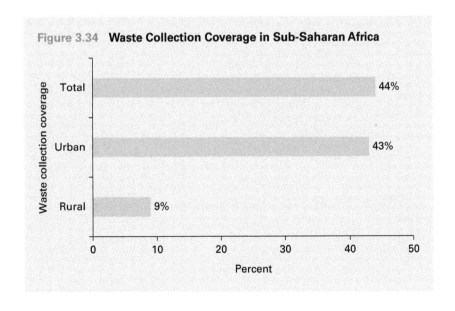

Figure 3.34 Waste Collection Coverage in Sub-Saharan Africa

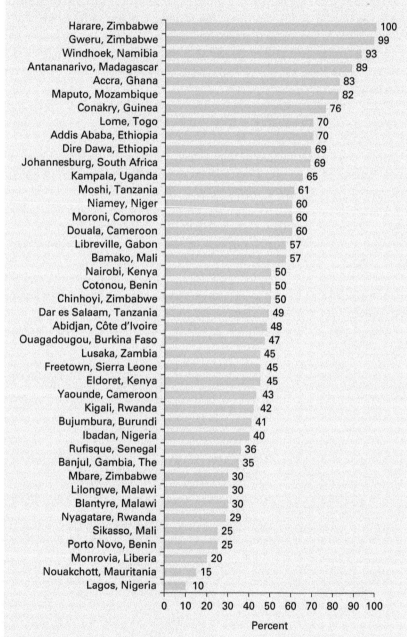

Figure 3.35 Waste Collection Rates for Select Cities in Sub-Saharan Africa
percent coverage

Note: Maximum chosen as reported for households, geographic area, or waste.

Photo 3.7 **Waste Collectors in Uganda**

are organized around high-income communities. Informal settlements and low-income areas often receive poor or no collection services because of social stigma, potential inaccessibility, violence and crime, and difficulties in fee collection.

Waste Transportation

Waste collection in the Sub-Saharan Africa region often occurs in two steps, with handcarts, tricycles, and donkeys commonly used to collect waste from households and for transportation to an aggregation site. From the aggregation site, small vehicles and trucks are used to bring waste to the final disposal site. However, some areas do not have a dual system and the waste is either dumped on empty land or in a canal, taken to a transfer station, or transported directly to the final disposal site. For formal urban waste disposal systems, the distance traveled from city center to final disposal sites can range from about 10 to 40 kilometers.

Waste Disposal

Currently, 69 percent of waste in the Sub-Saharan Africa region is openly dumped, and often burned (figure 3.36). Some 24 percent of waste is disposed of in some form of a landfill and about 7 percent of waste is recycled or recovered. As waste systems modernize, the amount of land-filling and recycling is anticipated to rise. More sanitary landfills are being built in the region; however, the number of new disposal facilities is not meeting the need given the growing quantities of waste. Because of

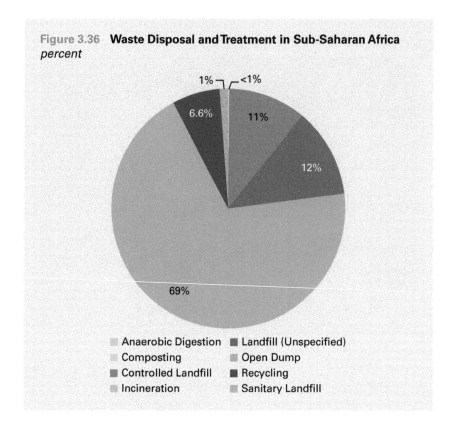

Figure 3.36   Waste Disposal and Treatment in Sub-Saharan Africa
percent

an increase in dumpsite failures affecting surrounding communities, the manner in which dumpsites are being operated is being improved and many are being closed completely. Cities are aware of recycling potential but recycling initiatives are most common in touristic cities.

Key challenges in the Sub-Saharan Africa region include overuse of facilities and continued disposal even after design capacity has been exceeded, citizen resistance to placement of waste facilities near their homes, land scarcity, and high urbanization and sprawl. Often, sufficient land is not available to service basic city functions, and governments are facing difficulty in coordinating services and investment at the intermunicipal level to save resources.

Waste Financing

Capital financing for necessary infrastructure investments alone is not a sufficient solution for Sub-Saharan Africa's solid waste sector. Many cities struggle with planning for long-term sustainability and with the financing of operational costs. In certain cases, cities have used donor funding to construct well-designed sanitary landfills that ultimately function as dumpsites because of lack of operational funding (World Bank 2017b).

In some cities, national and local governments are able to finance portions of the waste system, although general government funding and waste fee collection are typically not sufficient for waste operations. Partnerships with the private sector have been challenging for the region. Although international companies are interested in larger public-private partnership opportunities, there have been few successes. Financial, institutional, and political complications can make it difficult to attract international companies for waste activities. Municipalities are looking for ways to expand local capacity.

References

Asian Development Bank. 2013. "Solid Waste Management in Nepal: Current Status and Policy Recommendations." Asian Development Bank, Manila.

CCAC (Climate and Clean Air Coalition). n.d. Municipal Solid Waste Initiative, Solid Waste Management City Profile, São Paulo, Brazil; page 4.

City of Ottawa. 2018. "Solid Waste—Data and Reports." Ottawa, Ontario. https://ottawa.ca/en/residents/garbage-and-recycling/solid-waste-data-and-reports#facts-and-overview.

Croitoru, L., and M. Sarraf. 2017. "Le Cout de la Dégradation de l'Environnement au Maroc." Environment and Natural Resources Global Practice Discussion Paper No. 5, World Bank, Washington, DC.

Enayetullah, I., and Q. S. I. Hashmi. 2006. "Community Based Solid Waste Management through Public-Private-Community Partnerships: Experience of Waste Concern in Bangladesh." Paper presented at the 3R Asia Conference, Tokyo, Japan, October 30–November 1.

Government of Bermuda. n.d. "Tynes Bay Waste Treatment." Ministry of Works and Engineering. http://rossgo.com/Tynes%20Bay/Incinerator.html.

IFC (International Finance Corporation). 2014. *Handshake: Waste PPPs.* https://www.ifc.org/wps/wcm/connect/81efc00042bd63e5b01ebc0dc33b630b/Handshake12_WastePPPs.pdf?MOD=AJPERES.

India, Ministry of Urban Development. 2016. Swachh Bharat Mission (Collected from Swachh Surveskshan, 2017). Ministry of Housing and Urban Affairs. Government of India.

Li, Judy. 2015. "Ways Forward from China's Urban Waste Problem." The Nature of Cities website. https://www.thenatureofcities.com/2015/02/01/ways-forward-from-chinas-urban-waste-problem/.

Lizana, Juan. 2012. "Plan de Inclusión Social para Segregadores en las Provincias de Cusco, Calca y Urubamba." Programa de Desarrollo Regional PRODER Cusco. Cuadro N° 19: Número de segregadores por distrito del Cusco con un 25% de segregación en fuente.

Ministry of Environment, Forest and Climate Change. 2016. "Construction and Demolition Waste Management Rules." New Delhi. http://www.moef.gov.in/sites/default/files/C%20&%20D%20rules%202016.pdf.

Qatar General Secretariat for Development Planning. 2011. *Qatar National Development Strategy 2011–2016: Towards Qatar National Vision 2030*. Doha: Qatar General Secretariat for Development Planning. https://www.mdps.gov.qa/en/knowledge/HomePagePublications/Qatar _NDS_reprint_complete_lowres_16May.pdf.

Republic of Lebanon Ministry of Environment and UNDP. 2017. "Updated Master Plan for the Closure and Rehabilitation of Uncontrolled Dumpsites throughout the Country of Lebanon." Ministry of Environment, Beirut, and United Nations Development Programme, New York.

Seattle Public Utilities. 2016. "2015 Recycling Rate Report." Seattle, WA.

UN (United Nations). 2017. *World Population Prospects: The 2017 Revision*. New York: UN. https://www.un.org/development/desa/publications/world -population-prospects-the-2017-revision.html.

World Bank. 2016. "Rapid Assessment of Kabul Municipality's Solid Waste Management System." Report No. ACS19236, World Bank, Washington, DC.

World Bank. 2017a. "Maldives Clean Environment Project: Combined Project Information Documents / Integrated Safeguards Datasheet (PID/ ISDS)." World Bank, Washington, DC. http://documents.worldbank.org /curated/en/455971491655712283/pdf/ITM00194-P160739-04-08 -2017-1491655709183.pdf.

World Bank. 2017b. "Senegal Municipal Solid Waste Project. Project Information Document/Integrated Safeguards Data Sheet (PID/ISDS)." World Bank, Washington, DC. http://documents.worldbank.org/curated /en/581531500995135875/pdf/ITM00184-P161477-07-25-2017 -1500995132357.pdf.

Additional Resources

Europe Environment Agency. 2016. "Municipal Waste Management across European Countries." Copenhagen. https://www.eea.europa.eu/themes /waste/municipal-waste/municipal-waste-management-across -european-countries.

King, Neil. 2017. "The Rise of Waste-to-Energy in the GCC." Gulf Business website. http://gulfbusiness.com/rise-waste-energy-gcc/.

Mohee, Romeela, and Thokozani Simelane. 2015. *Future Directions of Municipal Solid Waste Management in Africa*. Pretoria: Africa Institute of South Africa.

Office of the Auditor-General of Papua New Guinea. 2010. "The Effectiveness of Solid Waste Management in Papua New Guinea." Performance Audit Report No. 01/2010. Waigani, Papua New Guinea.

Regional Environmental Center and Umweltbundesamt GmbH. 2018. "Waste Management Legislation." In *Handbook on the Implementation of EC Environmental Legislation.* Szentendre, Hungary: Regional Environmental Center; Dessau-Roßlau, Germany: Umweltbundesamt. http://ec.europa.eu/environment/archives/enlarg/handbook/waste.pdf.

Secretariat of the Pacific Regional Environment Programme. Waste characterization report 2014. https://www.sprep.org/j-prism/reports-a-materials.

United States Environmental Protection Agency. 2018. Municipal Solid Waste Landfills: National Emission Standards for Hazardous Air Pollutants (NESHAP). https://www.epa.gov/stationary-sources-air-pollution/municipal -solid-waste-landfills-national-emission-standards.

World Bank Open Data: Population 1960–2016. n.d. https://data.world bank.org/indicator/SP.POP.TOTL.

CHAPTER 4

Waste Administration and Operations

Key Insights

- In most countries, solid waste management is a local responsibility, by default or through decentralization policies. Direct central government involvement in waste services, other than regulatory oversight or subsidies, is rare.
- About 70 percent of waste services are overseen directly by local public entities, with the remainder administered through other levels of government, intermunicipal arrangements, mixed public-private entities, or private companies.
- About half of services are operated by public entities. About one-third of services, from collection to treatment and disposal of waste, are operated through mixed public-private partnerships.
- The private sector is typically engaged through management or concession contracts for collection, treatment, and disposal, and contracts generally last fewer than 10 years.
- Intermunicipal government cooperation is in place in a minority of cities, and typically occurs through the use of shared assets for waste transfer, disposal, and city cleaning.
- About two-thirds of countries have created targeted legislation and regulations for solid waste management, though enforcement may vary.
- Almost 70 percent of countries have established institutions with responsibilities for policy development and regulatory oversight in the waste sector.
- Systematic public reporting on waste policies and waste data is largely limited to high-income countries and some middle-income countries.

Waste management is an essential urban service that requires planning, management, and coordination across all levels of government and stakeholders. Solid waste management services typically include waste collection from households and commercial establishments and haulage to a collection point or transfer station, transportation from a collection point or transfer station to a final treatment or disposal site, treatment and disposal of waste, and street cleaning and drainage management. Countries and cities around the world are pursuing a range of administrative and operational models to offer some or all of these services.

In high-income countries, national governments develop laws and regulations that establish guidelines, national performance targets, and operational and environmental standards. In rare cases national governments may operate solid waste services, but solid waste management is typically a local service. Local governments, such as cities, prefectures, and states, are often responsible for creating more specific local regulations, collecting and disposing of waste, and deciding how physical and financial resources should be allocated and how costs can be recovered. Local agencies are also responsible for identifying private sector partners that may build or operate services, siting new landfills or other waste facilities, and monitoring service coverage, citizen feedback, and pollution from facilities. It is at the local level that innovative waste programs are typically developed, such as the introduction of bins of different colors for household source separation or local composting programs.

Adequate waste services are more difficult to achieve in low- and middle-income countries, where challenges are as much a result of poor planning and service operation as a lack of funding for investments. Daily waste management is expensive; requires institutional skills for planning, operational management, and oversight; and, where funding is limited, waste management competes with other development priorities. Developing waste management capacity and mobilizing resources requires strong political support. Typical challenges that have repeatedly been identified in World Bank studies include the following:

- Shortage of financial resources, particularly to operate waste collection, transport, and disposal systems, caused by lack of revenues from households and other waste generators or lack of budget and funding in local governments.
- Complexity of designing and managing decentralized, locally based waste collection, transport, and disposal systems while maximizing coverage and minimizing environmental impacts.
- Lack of land and resistance from local populations to development of waste facilities.
- Limited institutional capacity for planning, monitoring, and enforcement.
- Ambiguity around organizational structure and responsibility, and coordination both within the same level of government and between national, regional, and local governments.

This chapter presents findings at both national and local government levels regarding regulations, institutions, and practical approaches to meet these operational challenges.

Solid Waste Regulations

A foundational aspect of sustainable waste management is proper planning and oversight from central authorities. While waste management is typically a locally operated service, both national and local governments play a role in defining the regulatory framework within which solid waste management services can be developed, and this can affect private sector engagement. National governments are typically responsible for establishing environmental standards for waste management and for creating rules for fair and transparent procurement of services from the private sector. National laws encourage local governments to adhere to common social and environmental standards. Local governments also establish rules and regulations that guide households and institutions on the proper management and disposal of waste. Typically, the entity that regulates waste management is separate from the entity that operates services to promote accountability.

National Waste Regulation

In this study, 86 percent of countries and economies reported the existence of an official national law or guidelines that govern solid waste management. Table 4.1 summarizes the number of countries with established national laws or other guiding frameworks for solid waste management, which may be a specific piece of solid waste regulation or broader environmental and urban laws that address solid waste management.

Table 4.1 Existence of National Waste Management Regulation

Income group	Total number of countries	Number of countries with defined solid waste management laws or guidelines	Number of countries without defined solid waste regulations or guidelines	Number of countries with no information	Share of countries with defined solid waste management laws or guidelines (percent)
High income	78	75	2	1	96%
Upper-middle income	56	47	4	5	84%
Lower-middle income	53	47	1	5	89%
Low income	30	18	1	11	60%
All	217	187	8	22	86%

Low-income countries are most likely to lack specific laws on waste management; in general, solid waste management systems are more nascent in these countries. Waste management in lower-income countries is also often primarily addressed by municipalities or even communities.

The number of countries that have specific solid waste management laws increases significantly in middle-income countries. Some 88 percent of middle-income countries have guiding solid waste management laws or frameworks. The vast majority of countries without data on waste legislation are in Sub-Saharan Africa, where laws are still being developed.

Solid waste management laws range from broad environmental rules to targeted interventions. For example, Peru's General Law on Solid Waste Management (Ley General de Residuos Sólidos, 27314) addresses all aspects of solid waste management, from generation to final disposal. Many countries have formed specific laws that address specific waste streams, and often, legislation for municipal waste is separate from that for hazardous or medical waste.

Enforcement of laws is a common challenge. Enforcement requires adequate staffing, implementation of fees or other penalties, and cultural alignment with legislative goals. In Malaysia, the National Solid Waste Management Policy was created to standardize and improve waste management across the country (Wee et al. 2017). However, deployment efforts were challenged by limited financing, low staff technical capacity, and ambiguity in the policy's guidelines. In Rwanda, a national plastic bag ban was strictly enforced using border patrol guards to prevent illegal imports and multiple penalties for offenders, including fines, jail time, and public shaming (de Freytas-Tamura 2017).

Photo 4.1   **Plastic Bag Ban in Kenya**

Local Waste Regulation

Because responsibility for executing solid waste management systems typically falls on local governments, local rules and regulations are commonplace. Local regulations cover specific aspects of waste management, including source separation, household and commercial fees, disposal sites, bans on plastic or other materials, and the institutions and agencies that are responsible for implementing waste operations and initiatives.

This study reveals that most cities have some solid waste management rules and regulations (table 4.2). Out of the cities studied, 223 reported the presence of official solid waste management policies. Only 18 reported a lack of policies, and data were not available for 127 cities.

Solid Waste Planning

Solid waste strategies and plans at both the national and local levels allow agencies to comprehensively understand the current situation, identify future goals, and outline a detailed plan of action to advance the solid waste management sector. Planning allows all stakeholders—including different government agencies, citizens, associations, and the private sector—to be coordinated and allows investments to be made in an efficient and targeted manner.

National Strategies

In more advanced cases of waste governance, national governments may develop a five- to ten-year national strategy that details the current waste situation in the country and sets targets for the sector about recycling, financial sustainability, citizen awareness, promotion of a green economy,

Table 4.2 **Existence of Urban Waste Management Regulation**

Region	Number of cities with defined solid waste management rules and regulations	Number of cities without defined solid waste management rules and regulations	Number of cities without available information
East Asia and Pacific	32	0	8
Europe and Central Asia	51	6	34
Latin America and the Caribbean	20	5	14
Middle East and North Africa	19	0	10
North America	6	0	0
South Asia	74	6	3
Sub-Saharan Africa	21	1	58
All	**223**	**18**	**127**

reduction of greenhouse gases, and rehabilitation of contaminated sites. Sometimes the national strategy contains legally binding legislation or other guidelines for individuals and institutions. National governments may provide financing or technical expertise, such as by sharing costs or evaluating plans for the construction of new disposal sites, to help localities achieve national goals.

The Kenya National Environmental Management Authority published Kenya's *National Solid Waste Management Strategy* in 2014 in response to citizen complaints of poor waste management, outlining collective action mechanisms to systematically improve waste management (NEMA 2014; Akinyi 2016). Another example, in Mozambique, is a national strategy for integrated waste management that details the current waste management situation and outlines a 12-year framework for action to address the most pressing solid waste management challenges. The strategy provides guidance on topics from landfill construction to organization of waste pickers and outlines the roles of all stakeholders, including central governments, municipalities, businesses, waste pickers, residents, and nongovernmental organizations (Tas and Belon 2014). National strategies often define metrics, such as recycling rates, to track progress over time.

Local Master Plans

Because waste management is a local service, it is much more common for cities to develop a solid waste management–focused master plan than for countries to create a national strategy. Master plans formalize the locality's goals for solid waste management and plans for implementation. Solid waste master plans are comprehensive, outlining planned investments in infrastructure, citizen engagement strategies, environmental criteria and safeguards, and all aspects of waste collection, transport, and disposal.

Cities typically implement master plans at a more mature stage in the solid waste management sector. In this study, 130 cities reported the existence of a master plan, but plans are being implemented in only 87 of them (table 4.3).

Table 4.3   Existence and Implementation of Urban Master Plan

	Number of cities with integrated solid waste management plan	Number of cities where a master plan is being implemented
Yes	130	87
No	58	30
Unknown	180	251

In South Africa, the National Environmental Management: Waste Act of 2008 mandated all municipalities create an Integrated Waste Management Plan (City of Johannesburg 2011). The city of Johannesburg responded by creating a plan that details current waste generation and characteristics, disposal practices, key roles and responsibilities, and instruments for implementation, including funding sources. The plan also details goals and targets for waste minimization and recovery, information systems, pollution control, governance, and budgets.

Institutions and Coordination

Both national and local governments may establish departments dedicated to solid waste management, though formalized solid waste management institutions are more common at the local level. Centralizing solid waste management under a single entity can help ensure that planning processes are coordinated, resources are used efficiently, redundancy in function is avoided across departments, and all gaps in service are minimized and addressed comprehensively. Central agencies can also assume responsibility for enforcing solid waste policies and regulations.

At a macro level, 148 out of 217 countries and economies have national agencies dedicated to enforcing solid waste laws and regulations. Data were not available in 53 countries, and 16 countries do not have such an agency. Furthermore, 19 countries reported the existence of a dedicated solid waste management agency, authority, or department. However, more commonly, solid waste management falls under the purview of an institution with broader responsibilities, such as a ministry of environment, planning, or local government.

At a local level, departments dedicated to solid waste management are much more common (table 4.4). A department dedicated to solid waste management was reported by 216 cities; 21 cities lacked a dedicated department; and data were not available for 131 cities. In addition, 107 cities reported having dedicated government units that combat common solid waste management issues, such as illegal dumping and littering.

Coordination is required to ensure consistency between different levels of government. Many governments also struggle with overlaps in responsibilities across agencies or gaps in responsibilities, since activities related to solid waste management often cut across multiple departments. As a solution in Pakistan, a new Sindh Solid Waste Management Board was established to coordinate waste management decisions across agencies and municipalities (SSWMB 2017). Furthermore, a successful model of interjurisdictional coordination is detailed in case study 4 in chapter 7, highlighting a Japanese experience.

Table 4.4   **Oversight of Solid Waste Management in Cities**

Region	Number of cities with a department dedicated to solid waste management			Number of cities with a unit enforcing solid waste management issues such as dumping and littering		
	Yes	No	Unknown	Yes	No	Unknown
East Asia	38	0	2	21	0	19
Europe and Central Asia	51	9	31	23	7	61
Latin America and the Caribbean	20	3	16	0	0	39
Middle East and North Africa	17	1	11	9	1	19
North America	6	0	0	0	0	6
South Asia	66	6	11	51	8	24
Sub-Saharan Africa	18	2	60	3	2	75
All	216	21	131	107	18	243

Waste Management Operations

Waste management services are delivered in a variety of ways across the world (table 4.5). Waste management is most commonly managed by municipalities in a decentralized manner. Solid waste management programs are typically designed in response to local conditions such as financing availability, local norms, spatial layout of communities, and the ability of citizens to pay for services. Where local conditions allow, solid waste services can be managed on an intermunicipal scale. Intermunicipal cooperation is common in European Union countries such as France, Italy, and the Netherlands (Kolsut 2016), where coordination has led to economies of scale, cost savings through fewer investments and a wider span of financing sources, reduced staffing needs, and exchange of technical skills. Intermunicipal coordination is most effective when operational objectives and guidelines are similar across entities, such as in European Union states sharing stringent membership requirements and legislative frameworks. Tokyo, Japan, provides an example of intermunicipal coordination, where prefecture governments build final disposal sites shared by multiple municipalities. Coordination can be difficult between municipalities that have different environmental and waste management priorities, are geographically separated, or that manage waste across several disparate departments.

It is less common for national governments to operate solid waste management services. Mixing regulatory and operational responsibilities

Table 4.5 Examples of Waste Management Operations and Administrative Models

Country	Region	Administrative model	Operational model
Indonesia	East Asia and Pacific	Highly decentralized, responsibility of the municipality	Communities organize waste collectors with user fees, city organizes waste transport and disposal from local budget, private operators are typically not involved.
Cambodia	East Asia and Pacific	Decentralized waste collection and disposal operated through long-term concession to private sector	Private operator collects user fees and provides waste collection and disposal service.
Azerbaijan	Europe and Central Asia	Highly centralized national planning and oversight	Waste services are organized by regional administrations.

can lead to conflicts in accountability, and local governments are able to offer more efficient services and informed plans since they are directly connected to the demographic that they serve. However, national operation has taken place in limited cases. For example, in Azerbaijan, national administration of waste services has enabled streamlined reform, system modernization, and standardization of services across different regions. The country formed its national strategy in 2006, closed about 80 percent of its informal dumpsites, and expanded collection services from covering 60 percent of the population to covering more than 75 percent (Van Woerden 2016). Similarly, waste services in smaller countries, such as Jamaica and Malaysia, are centrally managed, though efforts are being made to separate responsibilities across different departments and to engage the private sector in service delivery (JSIF, n.d. Manaf, Samah, and Zukki 2009).

Slightly more than 30 percent of waste management services, from primary collection to treatment and disposal, are provided through public-private partnerships, even though such partnerships can be complex to structure and implement. Private operators can bring efficiency and financial security to waste management systems under the right conditions. Often, the private sector is engaged to help public entities improve operations and mitigate the common challenge of unequal collection access across economic cohorts. For example, in Lahore, Pakistan, a private company not only increased waste collection coverage from 51 percent to 88 percent but also enhanced waste monitoring through vehicle tracking, which improved the way that limited financial and physical resources were strategically used and allocated (Ashraf, Hameed, and Chaudhary 2016).

When private entities are involved, governments must consider the territory that the entity has control over. For example, cities have pursued varied models for waste collection. In the city of San Francisco in the United States, a single operator has a designated monopoly to provide waste collection services to the whole city, which encourages the company's participation in and ability to test new models and promotes consistency of services (Stevens 2011). In Singapore, however, the city is zoned so that different operators serve unique geographic sections, thus promoting more widespread participation and competition (Singapore National Environment Agency n.d.). Governments must balance competition against crowding of streets and revenue opportunities for private operators. Finally, some cities have commercial entities that directly contract with the hauler of their choice; however, this method can lead to other concerns, such as vehicle congestion.

In addition to general solid waste laws, 54 percent of countries studied have official laws guiding public entities in engagement with the private sector. For 21 percent of countries, such regulation is absent, and data were not available for 25 percent of countries. Effective public-private partnership laws provide institutional and financial security to private operators and increase the viability of engaging the private sector in solid waste management operations.

Cities use a variety of administrative and contracting models across the waste management value chain (figure 4.1). Almost all services are administered and operated by public entities, with 50–70 percent of cities reporting public administration at a municipal level and more than 40 percent reporting public operation from collection to disposal to treatment. Intermunicipal cooperation typically occurs for services involving shared assets, such as transfer stations or disposal sites, as well as for city cleaning, which may achieve economies of scale across shared spaces. The majority of cities with data reported that most waste services are operated publicly, but a small fraction, about 15 percent, are operated "directly," which means that a service is provided by independent small-scale organizations directly to households.

Where private entities are engaged, management, concession, and other public-private partnership contracts are most commonly used. Most contracts last less than 10 years, which provides flexibility to the public entity. For waste disposal, which typically entails the operation of a fixed asset such as a landfill, about 35 percent of contracts last 10 years or longer.

Figure 4.1 Waste Management Administration, Operation, and Financing Models

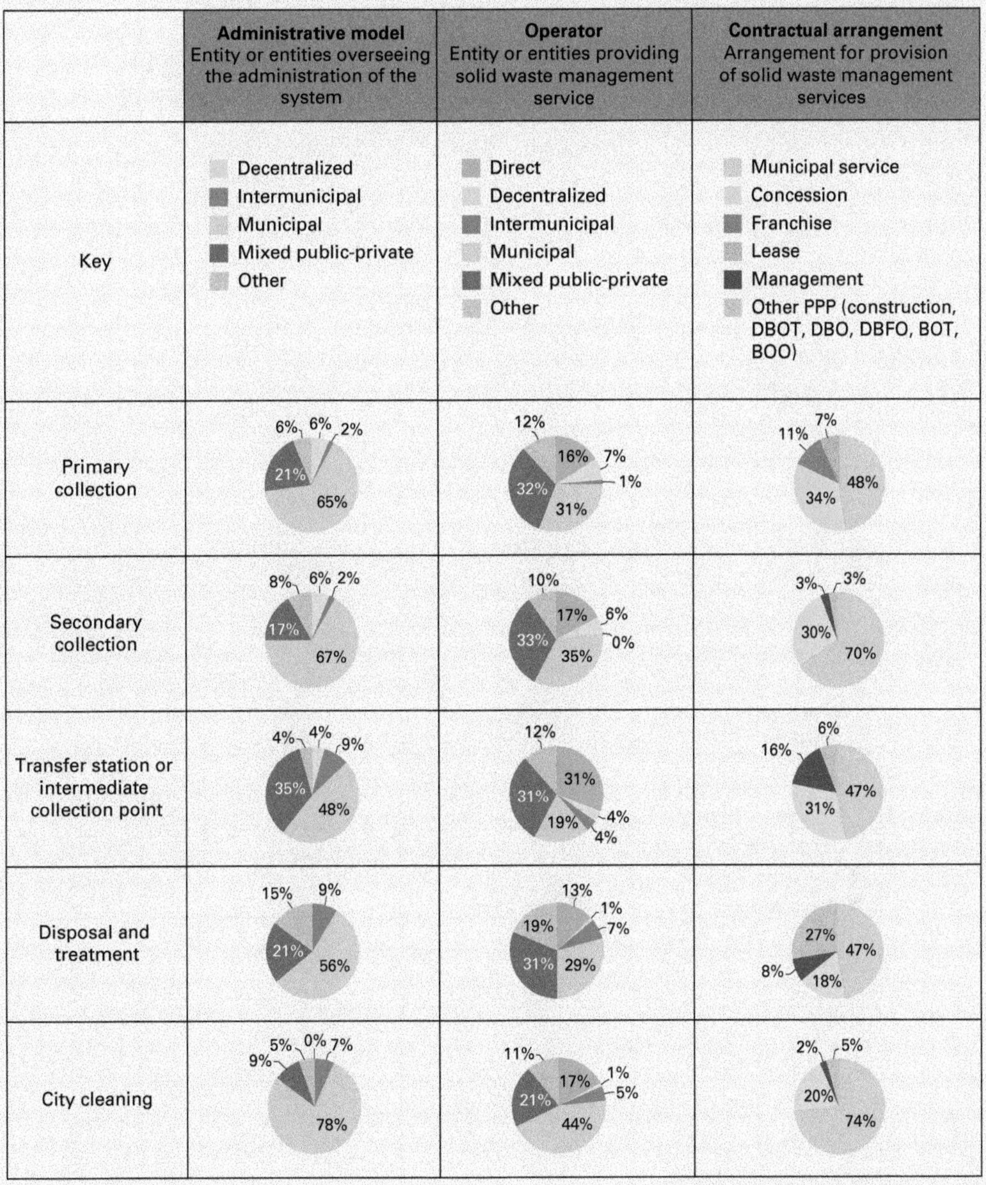

Administrative model
Entity or entities overseeing the administration of the system
Operator
Entity or entities providing solid waste management service
Contractual arrangement
Arrangement for provision of solid waste management services
Key
Decentralized
Intermunicipal
Municipal
Mixed public-private
Other
Direct
Decentralized
Intermunicipal
Municipal
Mixed public-private
Other
Municipal service
Concession
Franchise
Lease
Management
Other PPP (construction, DBOT, DBO, DBFO, BOT, BOO)
Primary collection
6% 6% 2% 21% 65%
12% 16% 7% 1% 32% 31%
7% 11% 48% 34%
Secondary collection
8% 6% 2% 17% 67%
10% 17% 6% 0% 33% 35%
3% 3% 30% 70%
Transfer station or intermediate collection point
4% 4% 9% 35% 48%
12% 31% 31% 19% 4% 4%
6% 16% 47% 31%
Disposal and treatment
15% 9% 21% 56%
13% 1% 19% 7% 31% 29%
27% 47% 8% 18%
City cleaning
5% 0% 7% 9% 78%
11% 17% 1% 21% 5% 44%
2% 5% 20% 74%

(Figure continues on next page)

Figure 4.1 Waste Management Administration, Operation, and Financing Models *(continued)*

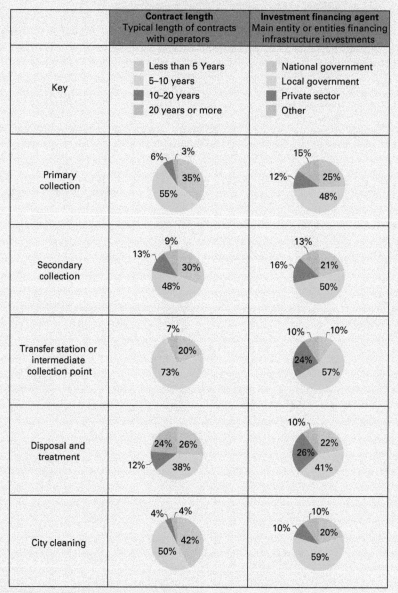

Note: **Decentralized** = local jurisdictions; **intermunicipal** = two or more municipalities in coordination; **municipal** = single municipality; **mixed public-private** = public and private entities; **other** = private, nongovernmental or other entity, or no formal administration system.
Direct = organization directly contracting with waste generator; **decentralized** = local jurisdictions; **intermunicipal** = two or more municipalities coordinate; **municipal** = single municipality; **mixed public-private** = public and private entities; **other** = private, nongovernmental or other entity, or none.
Municipal service = municipality provides services; **concession** = government grants private firm assets or opportunity to provide services in exchange for rights to profit; **franchise** = government contracts exclusively with private firm for long-term service provision in specific areas; **lease** = private operator pays municipality for use of public assets; **management** = government hires private operator to operate a waste facility; **construction** = contract for construction of facilities; **BOO** = build-operate-own; **BOT** = build-operate-transfer; **DBFO** = design-build-finance-operate; **DBO** = design-build-operate; **DBOT** = design-build-operate-transfer; **PPP** = public-private partnership.

References

Akinyi, Lucy. 2016. "NEMA Launches Solid Waste Management Strategy." Citizen Digital, September 15. https://citizentv.co.ke/news/nema-launches-solid-waste-management-strategy-141364/.

Ashraf, Usman, Isbah Hameed, and Muhammad Nawaz Chaudhary. 2016. "Solid Waste Management Practices under Public and Private Sector in Lahore, Pakistan." *Bulletin of Environment Studies* 1 (4): 98–105. http://www.mnpublishers.com/journals/bes/archives/issue/vol-1-no-4/article/solid-waste-management-practices-under-public-and-private-sector-in-lahore-pakistan.

City of Johannesburg. 2011. "City of Johannesburg Integrated Waste Management Plan." Johannesburg, South Africa. http://www.pikitup.co.za/wp-content/uploads/2015/10/City-of-Joburg-Integrated-Waste-Management-Plan-2011.pdf.

de Freytas-Tamura, Kimiko. 2017. "Public Shaming and Even Prison for Plastic Bag Use in Rwanda." *New York Times*, October 28. https://www.nytimes.com/2017/10/28/world/africa/rwanda-plastic-bags-banned.html.

JSIF (Jamaica Social Investment Fund). n.d. "National Solid Waste Management." http://www.jsif.org/content/national-solid-waste-management.

Kolsut, Bartłomiej. 2016. "Inter-Municipal Cooperation in Waste Management: The Case of Poland." *Quaestiones Geographicae* 35(2): 91–104.

Ley General de Residuos Sólidos. http://www.upch.edu.pe/faest/images/stories/upcyd/sgc-sae/normas-sae/Ley_27314_Ley_General_de_Residuos_Solidos.pdf.

Manaf, L.A., M.A. Samah, and N.I. Zukki. 2009. "Municipal Solid Waste Management in Malaysia: Practices and Challenges." *Waste Management* 29 (110): 2902–6. https://www.ncbi.nlm.nih.gov/pubmed/19540745.

NEMA (National Environment Management Authority, Kenya). 2014. *The National Solid Waste Management Strategy.* Nairobi: NEMA. https://www.nema.go.ke/images/Docs/Media%20centre/Publication/National%20Solid%20Waste%20Management%20Strategy%20.pdf.

Singapore National Environment Agency. n.d. "Waste Management: Overview." http://www.nea.gov.sg/energy-waste/waste-management.

SSWMB (Sindh Solid Waste Management Board). 2017. "Hands Together for Clean & Healthy Sindh." SSWMB, Sindh, Pakistan. http://www.urckarachi.org/downloads/SSWMB%20Forum%20by%20A.D.Sanjnani%2019%20April%202017.compressed.pdf.

Stevens, Elizabeth. 2011. "Picking Up the City's Garbage Is a Sweet Deal, and a Monopoly." *New York Times*, June 9. https://www.nytimes.com/2011/06/10/us/10bcstevens.html.

Tas, Adriaan, and Antoine Belon. 2014. *A Comprehensive Review of the Municipal Solid Waste Sector in Mozambique.* Nairobi: Carbon Africa. http://www.associacao-mocambicana-reciclagem.org/wp-content /uploads/2017/08/2014-08-05-A-Comprehensive-Review-of-the-Waste -Sector-in-Mozambique-FINAL.pdf.

Van Woerden, Frank. 2016. "VPU Team Awards for ECA: Azerbaijan Integrated Solid Waste Management Project." Washington, DC, World Bank.

Wee, Seow Ta, Muhamad Azahar Abas, Sulzakimin Mohamed, Goh Kai Chen, and Rozlin Zainal. 2017. "Good Governance in National Solid Waste Management Policy (NSWMP) Implementation: A Case Study of Malaysia." *AIP Conference Proceedings* 1891. https://aip.scitation.org /doi/pdf/10.1063/1.5005461.

CHAPTER 5

Financing and Cost Recovery for Waste Management Systems

Key Insights

- Basic solid waste management systems covering collection, transport, and sanitary disposal in low-income countries cost $35 per tonne at a minimum and often much more.
- Solid waste management is a large expenditure item for cities and typically comprises nearly 20 percent of municipal budgets in low-income countries, more than 10 percent in middle-income countries, and 4 percent in high-income countries. Budgets can be much higher in certain cases.
- Systems that include more advanced approaches for waste treatment and recycling cost more, from $50 to $100 per tonne or more. The choice of waste management methodology and technology depends highly on the local context and capacity for investments and ongoing management.
- User fees range from an average of $35 per year in low-income countries to $170 per year in high-income countries. Full cost recovery from user fees is largely limited to high-income countries. Almost all low-income countries, and a limited number of high-income countries, such as the Republic of Korea and Japan, subsidize domestic waste management from national or local budgets.
- Although public-private partnerships could potentially reduce the burden on local government budgets, they could result in compromises in service quality when not structured and managed properly.
- Local governments provide about 50 percent of investments for waste services, and the remainder is typically provided through national government subsidies and the private sector.
- When political support for increasing user fees for households to cost recovery levels is limited, cross-subsidizing from payments by waste generators (for example, the commercial sector) can help reduce the burden on local government budgets. Commercial fees range from about $150 per year in low-income countries to $300 in high-income countries.
- Volume-based waste fees have been successful in countries like Austria, Korea, and the Netherlands but are still uncommon because they require coordinated planning and strong enforcement. Households and commercial institutions in low-income countries are typically charged a flat fee that is collected on a door-to-door basis.

Because solid waste management is commonly a locally managed service, operations and financing often fall under the purview of local governments. In low- and middle-income countries, waste management financing is often limited, and funding must be balanced with the provision of other essential services, such as health care, education, and housing. Given the potential major impact of the financial sustainability of waste systems on the overall health of the city, designing an efficient system with clear paths for financing is essential. Furthermore, a well-functioning system can create a positive feedback loop in which citizens gain trust and satisfaction with services and are more willing to pay. With a growing global economy and a global population that is anticipated to increase from 7.6 billion today to 9.8 billion in 2050, the importance of financial efficiency in solid waste management has never been greater (United Nations 2017).

Waste Management Budgets

Waste management is an expensive service and requires substantial investments in physical infrastructure and long-term operations. Solid waste management services are also essential to the physical and economic health of society and are often a priority budget item for cities. For cities in low-income countries, solid waste management expenditures, on average, comprise 19 percent of municipal budgets (table 5.1). As countries grow economically, more funding is allocated to other public services.

Despite the substantial share of solid waste management expenditures in municipal budgets, low- and middle-income countries often face budget shortfalls for waste services and thus reduction of costs and recovery of fees is often integral to the development of the sector.

Table 5.1 Solid Waste Management as a Percentage of Municipal Budget

Income group	Average percentage of municipal expenditures on solid waste management
High income	4%
Middle income	11%
Low income	19%

Note: The absolute average of municipal expenditures on solid waste management was used. Only one city per country is represented in this analysis to prevent skewing, for a total of 46 countries. The capital city was selected if data were available, otherwise the next largest city was used. When data from multiple cities were available, budget ratios were found to be similar across cities within a single country.

Waste Management Costs

Municipalities providing waste management services generally experience two broad categories of expenditures: (1) capital expenditures, which are typically associated with infrastructure investments; and (2) operational expenditures, often associated with service provision and equipment maintenance. Planning around these two types of expenditures generally differs.

The largest one-off waste management expenditure for municipalities is typically for infrastructure investment. Construction of sanitary disposal sites and purchase of collection and disposal equipment and bins is a prerequisite to offering consistent services to residents. The cost of construction and maintenance of disposal facilities may influence the city's choice of final disposal strategies. For example, landfill construction can cost a municipality roughly US$10 million to serve a population of 1 million people; the cost of a composting facility can range from a few million dollars for basic (windrow) composting facilities to about US$10 million for highly mechanized outfits; an incinerator with heat and energy recovery cost about $600/annual tonne for capital costs (defined as the total capital cost for the lifetime of the plant divided by the total annual capacity) for recent plants in Mexico, Poland, Singapore, and the United States (Kaza and Bhada-Tata 2018). Transfer stations can be very basic with a cost of about US$500,000, but when recycling and sorting functions are included, investment increases by several times. The technology that is most feasible depends not only on financial stability, but also on the technical capacity and local environment of the city.

Cities offering collection services must also purchase vehicles. In middle- and high-income countries, large new trucks cost about US$250,000 each, while low-income countries often use more localized systems that minimize investment costs, such as buggies, handcarts, and donkeys (Lee 2009). Cities must balance the fact that newer vehicles are more fuel efficient and require fewer parts and less maintenance, but have high initial investment costs. Along with capital costs, cities must also factor in the cost of feasibility studies and environmental and social assessments that take place ahead of construction projections.

The largest financial challenge for cities is usually the coverage of operational expenditures for labor, fuel, and the servicing of equipment. For example, for the city of Istanbul, Turkey, labor costs account for 58 percent of operational costs for the public collection system, and fuel accounts for another 31 percent (Dogan and Suleyman 2003). Tipping fees may be a source of revenue or expenditure depending on whether a local government is paying a private disposal facility or private haulers and residents are paying to use a municipally operated site. Governments that operate waste management systems must also factor in costs of repairs, depreciation of vehicles and other assets, operational costs for landfill operation such as a daily cover, and utilities and overhead costs. For example, in Bahir Dar, Ethiopia, equipment was assumed to depreciate at a rate of 20 percent a year (Lohri, Camenzind, and Zurbrügg 2014). In Seattle in the United States, overhead costs comprised 22 percent of the total waste management operating budget (DSM Environmental Services 2012).

Operating costs are almost always substantially higher than capital costs for investments and are often the most challenging to sustain. Even when capital costs are accounted for (often funded separately, for example, with national government subsidies), operational expenditures can easily account for 70 percent or more of total required budgets. Across collection and disposal operations, waste collection typically accounts for 60–70 percent of total costs. However, disposal costs have risen with more advanced sorting and materials recovery choices. For many cities, the long-term environmental benefit of operational expenditures, including the availability of raw materials and preservation of land value, outweighs the higher initial costs. See table 5.2 for a summary of typical waste management expenditures across major categories.

Disposal costs vary greatly. In some countries, waste disposal is informal and therefore not officially accounted for. In high-income countries, disposal costs are better accounted for and are typically between US$50 and US$100 per tonne. Recycling costs for high-income countries are often comparable to the cost of landfilling. Recycling is sometimes made cheaper when landfills are taxed or when limited capacity is available and market prices for landfilling increase. Construction and operation of anaerobic digestion and incineration systems require a large budget (table 5.3) and high management and technical capacity, and the technologies are rarely used for municipal waste in low- and middle-income countries.

The cost of open dumping is difficult to quantify because of a lack of data on construction and tipping fees. However, dumping incurs substantial costs in lost land value and increases the risk of high disaster-related

Table 5.2   **Typical Waste Management Costs by Disposal Type**
US$/tonne

	Low-income countries	Lower-middle-income countries	Upper-middle-income countries	High-income countries
Collection and transfer	20–50	30–75	50–100	90–200
Controlled landfill to sanitary landfill	10–20	15–40	20–65	40–100
Open dumping	2–8	3–10	—	—
Recycling	0–25	5–30	5–50	30–80
Composting	5–30	10–40	20–75	35–90

Source: World Bank Solid Waste Community of Practice and Climate and Clean Air Coalition.
Note: — = not available.

Table 5.3   Capital and Operational Expenditures of Incineration and Anaerobic Digestion Systems

US$/tonne

	Incineration		Anaerobic Digestion	
	Capital Expenditures[a] (US$/annual tonne)	Operational Expenditures (US$/tonne)[b,c]	Capital Expenditures (US$/annual tonne)	Operational Expenditures (US$/tonne)
Europe	$600–1,000	$25–30	$345–600	$31–57
United States	$600–830	$44–55	$220–660	$22–55
China	$190–400	$12–22	$325	$25

Source: Kaza and Bhada-Tata 2018.
Note: MWh = megawatt hour of energy.
a. In Europe and the United States, predominantly mass-burn/moving grate technology is used for waste incinerators with energy recovery (waste-to-energy incineration). In China, many incinerators use circulating fluidized bed (CFB) technology, which reflects the lower end of investment cost, although moving grate incinerators are also becoming more common.
b. Operating costs without accounting for revenues range between $100-200/tonne. The figures presented in the table are typical operating costs (net gate fees) taking into account revenues for electricity and heat sales and other revenues. In Europe, also including subsidies to energy from waste in some countries, these revenues are typically about $100/tonne, hence the resulting operating costs. In the United States, feed-in tariffs for electricity are typically lower, below $50/MWh.
c. Mixed waste in the United States and Europe is relatively low in organics and water content and hence high in calorific value. As a consequence, operating costs for waste with high organics often seen in lower-income countries could substantially increase operating costs due to lower revenues.

costs depending on the proximity and density of the local population to the disposal site. Dumpsite closures can also result in significant costs. In addition to the costs of land, disasters, and dump closures, poor waste management using dumping or uncontrolled burning results in environmental costs from air and water pollution and damage to human health. These economic costs can often be significant in the long run.

Waste Management Financing

Financing waste management systems is often one of the greatest concerns for municipalities. Cost recovery is essential to avoid reliance on subsidies from own-source revenues or from national or external sources.

Waste management investment costs and operational costs are typically financed differently. Given the high costs associated with infrastructure and equipment investments, capital expenditures are typically supported by subsidies or donations from the national government or international donors, or through partnerships with private companies.

About half of investments in waste services globally are made by local governments, with 20 percent subsidized by national governments, and 10–25 percent from the private sector, depending on the service provided.

In a public-private partnership in Siam Reap, Cambodia, most waste collection and some waste disposal is contracted to private operators without public budget support (Denney 2016). The private operators directly collect user fees for their services to cover expenses.

Operational expenditures typically require a solid cost-recovery system for long-term sustainability. The starting point for many municipalities is a standard user fee, which is charged to users for services delivered. User fees may be fixed or variable to encourage reduced waste generation or to provide affordability for lower-income residents. The most effective user fees match the ability and willingness of users to pay. For example, in a single city, poor neighborhoods may be charged no fees while more wealthy neighborhoods or those serviced by the private sector or a nongovernmental organization pay regular fees. In Yunnan, China, for instance, households in urban areas pay US$1.5 per month for waste services, whereas services are offered to rural areas at no cost (Zhao and Ren 2017). User fees may be billed through an independent waste service bill or in combination with other utility and property taxes to increase fee recovery. In addition to user fees, cities may recover costs by selling recycled materials and compost, generating energy from waste, establishing a financial deposit system on recyclables such as water bottles, taxing consumer goods such as plastic bags and batteries, and levying licensing fees from operators of transfer stations and final disposal sites.

Table 5.4 presents averages across regions, showing that household fees vary greatly. User fees in the Europe and Central Asia region were found to be the highest and those in Sub-Saharan Africa were the lowest.

Fees also vary greatly by income level, with residents in high-income countries paying substantially higher fees for services than paid by residents in lower-income countries. For low-income countries, fees are usually a flat amount per household. Volume-based fees are common in higher-income countries. Joint billing with property or utility taxes is practiced for middle- and high-income countries. Joint billing requires significant coordination and therefore maturity of waste management systems, but leads to higher

Table 5.4   Waste Management User Fees by Region

Region	Average user fee in selected cities (US$/year, as reported in data)
East Asia and Pacific	46
Europe and Central Asia	83
Latin America and the Caribbean	80
Middle East and North Africa	55
South Asia	34
Sub-Saharan Africa	10–40 (based on World Bank estimates)

Table 5.5   Waste Management User Fees by Income Level

Income group	Average fees, US$ per year	
	Household	Commercial
High income	$168	$314
Upper-middle income	$52	$235
Lower-middle income	$47	$173
Low income	$37	$155

Note: All currency amounts are in US$.

cost recovery amounts. In lower-middle-income countries and low-income countries, fees are more often collected door to door.

Fees charged to commercial institutions often vary by tonnage produced; fees are highest in high-income countries, where annual commercial fees average US$314 (table 5.5). For some high- and middle-income countries, fees are flat for each business, which is simpler to administer and collect. Low-income countries tend to charge fees for waste services less often, and data availability is scarce.

Figure 5.1 shows distributions of fee types and billing methods across countries by income categories.

In most countries, the cost of integral waste services (collection, transport, treatment, and disposal) cannot be fully recovered from user fees and requires subsidies through government transfers or external budget support. According to the What a Waste 2.0 study, local governments that receive transfers or subsidies for solid waste programs typically receive between US$4 and US$10 per capita per year. The average of subsidies or transfers from central governments is US$8 per capita per year. The agency providing funding may be the national government or a regional government. For example, in Sarajevo, Bosnia and Herzegovina, the cantonal government provided almost US$4.5 million for intermunicipal communal services such as street cleaning (KJKP 2016). In Yangon, Myanmar, the regional government funds city waste services, including salaries, uniforms, and the purchase and maintenance of equipment, which amounted to US$8.2 million in 2014 (CCAC, n.d.). In Majuro, Marshall Islands, the city government receives an annual operating subsidy of $325,000 from the national government (ADB 2014).

A partnership with the private sector is commonly pursued as a mechanism for achieving efficiency, technical expertise, and financial investment in waste management systems. In Istanbul, Turkey, waste collection was found to be 38 percent more cost efficient when operated by a private operator rather than a public operator (Dogan and Suleyman 2003). Private corporations may participate at all steps in a waste management value chain, including construction and operation of disposal sites and transfer

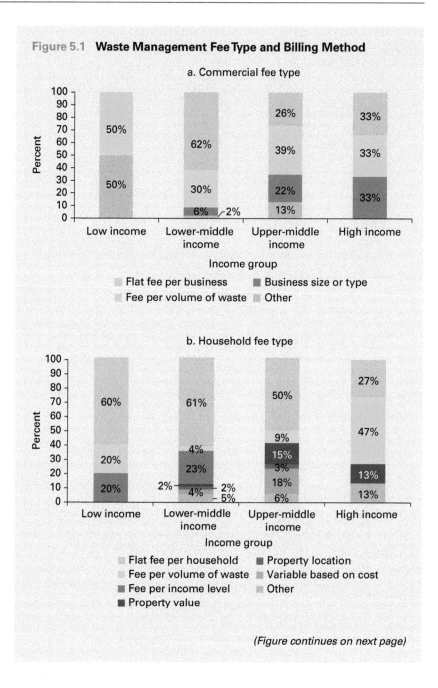

Figure 5.1   Waste Management Fee Type and Billing Method

(Figure continues on next page)

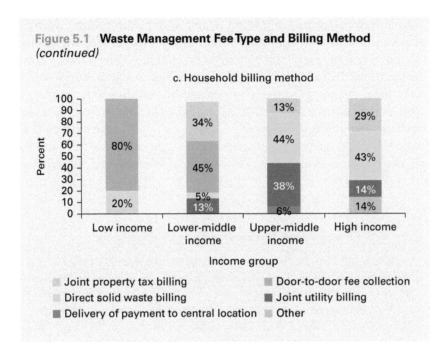

Figure 5.1 Waste Management Fee Type and Billing Method *(continued)*

c. Household billing method

Joint property tax billing

Direct solid waste billing

Delivery of payment to central location

Door-to-door fee collection

Joint utility billing

Other

stations, waste collection from homes and businesses, street cleaning, and citizen education on waste reduction and source separation. Private partners recover costs through their service provision. Therefore, successful municipalities ensure that private corporations are either paid directly by the locality or are provided with stable opportunities to earn revenues from tipping fees, user fees, or the sale of recycled materials. Environments that are typically conducive to private sector partnerships include simple and transparent procurement processes, minimal political and currency risk, and strong legal systems that enforce payments and encourage user compliance with waste management rules and regulations, such as those about littering and source separation. The lower the risks, the more likely it is for a private corporation to participate in the waste management system.

A unique form of private sector participation is the extended producer responsibility (EPR) system. In an EPR system, the cost for the final recycling or disposal of materials is borne by the producer of the good. Producers may pay the municipality directly for the cost of collection and disposal or develop a system for citizens to return the product. In either case, producers will often price the cost of disposal into the product so that consumers ultimately bear the disposal cost. Therefore, both producers and consumers are financially and logistically responsible for their resource usage. EPR systems ultimately reduce government costs, divert waste from disposal facilities to save space, and encourage environmentally friendly consumption (Product Stewardship Institute 2014).

Solid waste management systems in low- and middle-income countries that are in early development or undergoing expansion often pursue external financing, especially for capital expenditures. Where waste management initiatives are aligned with national objectives, local governments may obtain financing from national transfers. Local waste management projects may also be financed through loans and grants from development agencies or regional banks that also commonly provide technical project support. Some financiers are testing a model in which payments are tied to proven outcomes. This model of results-based financing is detailed in box 5.1.

Box 5.1   Results-Based Financing in Waste Management

An increasingly common strategy for promoting the efficient use of limited funds and for creating sustained behavior change is results-based financing (RBF). In this financing structure, payment for solid waste services depends on the delivery of predetermined results (Lee et al. 2014). By tying financing to outcomes, RBF encourages stakeholders to operate efficiently and change their behavior. RBF from governments or external institutions can be tailored to achieve several objectives and help waste management stakeholders do the following:

- *Increase fee collection,* such as by matching a portion of the fees collected by the managing institution
- *Promote source separation, waste reduction, and recycling,* such as by providing a stipend to neighborhoods that sort and separate an adequate quantity of clean recyclables
- *Strengthen waste collection and transportation,* such as by paying waste collectors upon successful and timely delivery of waste to the final disposal site
- *Design efficient infrastructure projects,* such as by making loans or grants for a new landfill project contingent on successful construction of various phases
- *Defray risk for investors and increase investments,* such as by delaying payments until proof of service success or completion of infrastructure

In Nepal, the World Bank supported a project to help bridge the gap between the costs of delivering improved waste management services and the revenues gained from user fees. For this project, payments were made based on the achievement of benchmarks such as the number of households receiving daily waste collection services, the cleanliness of public areas, and the feedback of households. These interventions greatly improved service quality. Furthermore, the RBF approach helped municipalities to gain financial stability by sustainably increasing user fees and improving the recovery of user fees by providing subsidies in proportion to the amount of fees collected by cities.

Another example is a World Bank-supported project in the West Bank, where results-based financing was used to combine the financing of the service to the poor while leveraging private sector engagement in managing and operating the sanitary landfill, transfer stations, and transportation of waste.

Finally, carbon financing is a strategy that has been used in limited cases by waste management projects that reduce greenhouse gas emissions, such as the installation of landfill gas capture infrastructure or composting of organic waste (box 5.2).

The appropriate sources of financing depend heavily on the local context, and a mix of strategies is often used to sustainably implement a solid waste management project.

Box 5.2   Carbon Finance

Carbon finance provides payments to projects that reduce greenhouse gas emissions by allowing entities that can reduce emissions at a low cost to receive payments from entities whose costs of reducing emissions are high. The entity that buys emissions reduction credits can claim credit for reducing greenhouse gas emissions, although it did not implement the project itself. Purchasing emissions credits can help countries achieve national climate goals, such as a Nationally Determined Contribution created as part of the international Paris Climate Agreement, and allow private entities to adhere to national climate laws or offset emissions from business activities (Müller 2017; World Bank 2015).

Solid waste projects that avoid greenhouse gas emissions include the capture of methane gas at a landfill or the composting of organic waste, as compared to the higher release of methane from decomposing waste at a dump or landfill. Waste projects that reduce emissions can sell certified emissions reductions on a public or private emissions trading system. The European Union, countries such as New Zealand and Switzerland, and states and provinces such as California, United States, and Ontario, Canada, have established emissions trading systems (UNDP 2017). For example, several landfill gas capture projects in Brazil were partially financed by selling emissions reductions through the World Bank's Carbon Partnership Facility (World Bank 2014, 2018).

A historical framework that motivated carbon trading was the Kyoto Protocol, established at the United Nations Framework Convention on Climate Change (UNFCCC) in 2005, through which industrial countries committed to emissions reduction targets (UNEP n.d.). Through the UNFCCC's Clean Development Mechanism (CDM), wealthy countries could invest in carbon-reducing projects in low- and middle-income countries to meet their Kyoto commitments (Appasamy and Nelliyat 2007). Through the CDM, several waste projects, such as composting plants in Bangladesh and Uganda, were implemented (Waste Concern 2014; AENOR 2009). However, since the commitments to the Kyoto Protocol expired in 2012, carbon trading for waste projects has greatly waned. Today, carbon financing for solid waste management faces mixed results and depends heavily on regulatory frameworks regarding emissions and air quality to establish an active marketplace with high prices.

A more recent form of carbon finance is the Pilot Auction Facility for Methane and Climate Change Mitigation, which is an innovative mechanism that encourages private sector investments in projects that reduce emissions.

References

ADB (Asian Development Bank). 2014. "Solid Waste Management in the Pacific: The Marshall Islands Country Snapshot." ADB Publications Stock No. ARM146608-2. Asian Development Bank, Manila. https://www.adb.org/sites/default/files/publication/42669/solid-waste-management-marshall-islands.pdf.

AENOR (Spanish Association for Standardization and Certification). 2009. "Validation of the Program of Activities: Uganda Municipal Waste Compost Programme." https://cdm.unfccc.int/ProgrammeOfActivities/poa_db/JL4B8R2DKF90NE6YXCVOQ3MWSGT5UA/V7Q8JRD1BKHUZ9S4L2OMIG5NFP6XY0/ReviewInitialComments/GLDOBAGRHOJT4IVWRZCTN8UV6KSJT4.

Appasamy, Paul P., and Prakash Nelliyat. 2007. "Financing Solid Waste Management: Issues and Options." *Proceedings of the International Conference on Sustainable Solid Waste Management, 5–7 September 2007, Chennai, India*, 537–42. https://www.researchgate.net/publication/237613599_Financing_Solid_Waste_Management_Issues_and_Options.

CCAC (Climate and Clean Air Coalition). n.d. "Solid Waste Management City Profile: Yangon, Myanmar." Accessed April 11, 2017. http://www.waste.ccacoalition.org/sites/default/files/files/yangon_city_profile_final_draft.pdf.

Denney, Lisa. 2016. "Reforming Solid Waste Management in Phnom Penh." Asia Foundation, San Francisco, CA; and Overseas Development Institute, London. https://asiafoundation.org/wp-content/uploads/2016/06/Working-Politically-and-Flexibly-to-Reform-Solid-Waste-Management-in-Phnom-Penh.pdf.

Dogan, Karadag, and Sakar Suleyman. 2003. "Cost and Financing of Municipal Solid Waste Collection Services in Istanbul." *Waste Management and Research* 21 (5): 480–85. http://journals.sagepub.com/doi/abs/10.1177/0734242X0302100511.

DSM Environmental Services. 2012. "Solid Waste Management and Municipal Finance." Presentation prepared for Connecticut Governor's Recycling Working Group, July 24. http://www.ct.gov/deep/lib/deep/waste_management_and_disposal/solid_waste/transforming_matls_mgmt/gov_recycling_work_group/appendix_d.pdf.

Kaza, Silpa, and Perinaz Bhada-Tata. 2018. "Decision Maker's Guides for Solid Waste Management Technologies." World Bank Urban Development Series, World Bank, Washington, DC.

KJKP 'RAD' d.o.o. Sarajevo. 2016. "Financial Operations Report for Cantonal Public Utility 'Rad' in Sarajevo for the period of January–December 2015" ["Izvještaj o Finansijskom Poslovanju KJKP 'RAD' d.o.o. Sarajevo za Period I–XII 2015. Godine"]. KJKP 'RAD' d.o.o. Sarajevo. Sarajevo, February.

Lee, Jennifer. 2009. "Sanitation Dept. Unveils Hybrid Garbage Trucks." *New York Times*, August 25. https://cityroom.blogs.nytimes.com/2009 /08/25/sanitation-dept-unveils-hybrid-garbage-trucks/.

Lee, Marcus, Farouk Banna, Renee Ho, Perinaz Bhada-Tata, and Silpa Kaza. 2014. *Results-Based Financing for Municipal Solid Waste.* Washington, DC: World Bank.

Lohri, C., E. J. Camenzind, and C. Zurbrügg. 2014. "Financial Sustainability in Municipal Solid Waste Management—Costs and Revenues in Bahir Dar, Ethiopia. *Waste Management* 34 (2): 542–52. https://www.science direct.com/science/article/pii/S0956053X1300500X.

Müller, N. 2017. "CER Demand, CDM Outlook and Article 6 of the Paris Agreement." CDM Training Workshop for DNAs and Stakeholders in Pakistan, Islamabad, Pakistan, August 21–22. https://unfccc.int/files/na /application/pdf/04_current_cer_demand_cdm_and_art__6_of_the_pa _nm.pdf.

Product Stewardship Initiative. 2014. "Electronics EPR: A Case Study of State Programs in the United States." Product Stewardship Initiative, Boston, MA. https://www.oecd.org/environment/waste/United%20States%20 (PSI%20-%20Cassel).pdf.

UNDP (United Nations Development Programme). 2017. "Carbon Markets." United Nations Development Programme, New York, NY. http://www.undp.org/content/sdfinance/en/home/solutions/carbon -markets.html.

UNEP (United Nations Environment Programme). n.d. "Clean Development Mechanism." UNEP Collaborating Centre on Energy and Environment, Risø National Laboratory, Roskilde, Denmark. https://unfccc.int/files /cooperation_and_support/capacity_building/application/pdf/unepcd mintro.pdf.

United Nations. 2017. "World Population Projected to Reach 9.8 Billion in 2050, and 11.2 Billion in 2100." United Nations, New York. https:// www.un.org/development/desa/en/news/population/world-population -prospects-2017.html.

Waste Concern. 2014. "22,783 CERs (Carbon Credits) Issued by UNFCCC for Recycling of Organic Waste in Dhaka, Bangladesh." Waste Concern, Bhaka, Bangladesh. http://wasteconcern.org/22783-cers-carbon-credits -issued-by-unfccc-for-recycling-of-organic-waste-in-dhaka-bangladesh/.

World Bank. 2015. "Private Sector—An Integral Part of Climate Action Post-Paris." World Bank, Washington, DC. http://www.worldbank.org /en/news/feature/2015/12/30/private-sector-an-integral-part-of -climate-action-post-paris.

———. 2014. "World Bank Carbon Funds and Facilities." World Bank, Washington, DC. http://www.worldbank.org/en/topic/climatechange /brief/world-bank-carbon-funds-facilities.

————. 2018. "Brazil—Integrated Solid Waste Management and Carbon Finance Project." Independent Evaluation Group, Project Performance Assessment Report 123798, World Bank, Washington, DC. http://documents .worldbank.org/curated/en/395271521557013485/pdf/123798-PPAR -P106702-P124663-P164310-PUBLIC.pdf.

Zhao, Zhen, and Xin Ren. 2017. "Yunnan Municipal Solid Waste Management and Implementation." Presented at World Bank Solid Waste Management Technical Deep Dive. Tokyo, Japan, March 21–24.

CHAPTER 6

Waste and Society

Key Insights

- Uncollected waste and poorly disposed of waste significantly affect public health and the environment, with the long-term economic impact of environmental recovery often resulting in multiple times the costs of developing and operating simple, adequate waste management systems.
- Waste management contributes nearly 5 percent of global greenhouse gas emissions, mainly driven by food waste and improper management of waste. Even basic system improvements can reduce these emissions by 25 percent and more.
- Of the cities with available data, 29 reported having completed an environmental assessment in the past five years while 73 cities reported no recent formal environmental assessment.
- Waste management systems should take into account potential extreme weather such as heavy storms that may cause the collapse of formal or informal waste facilities or damage urban infrastructure.
- High-income countries are attempting to divert waste from landfills and incinerators and increase adoption of recycling and waste reduction.
- More than 15 million people globally earn a living informally in the waste sector (Medina 2010). Waste pickers—often women, children, the elderly, the unemployed, and/or migrants—are a vulnerable demographic.
- The number of female waste pickers can often exceed the number of male waste pickers. In Vientiane, Lao PDR, and Cusco, Peru, 50 percent and 80 percent of waste pickers are female, respectively (Arenas Lizana 2012; Keohanam 2017).
- Of the cities with available data, 24 reported having completed a social assessment in the past five years while 73 cities reported no recent formal social assessment.

Waste management has broad societal impacts. The way that waste is managed affects the health of the environment, the livelihood and well-being of vulnerable populations, and the relationships between governments and citizens. Solid waste management influences how a society lives on a daily basis, and its strengths and failures can have a magnified impact during crisis situations. Although limited data were collected for this report regarding environment and climate change, citizen engagement, and the informal sector, the report discusses these key topics, in addition to technology trends, to highlight several critical aspects that a well-functioning waste management system should consider.

Environment and Climate Change

Solid waste management is inextricably linked to environmental outcomes and their subsequent economic consequences. At the local and regional levels, inadequate waste collection, improper disposal, and inappropriate siting of facilities can have negative impacts on environmental and public health. At a global scale, solid waste contributes to climate change and is one of the largest sources of pollution in oceans.

In low- and many middle-income countries, inadequate waste collection and uncontrolled dumping or burning of solid waste are still an unfortunate reality, polluting the air, water, and soil. When waste is burned, the resulting toxins and particulate matter in the air can cause respiratory and neurological diseases, among others (Thompson 2014). Piles of waste produce toxic liquid runoff called leachate, which can drain into rivers, groundwater, and soil. Organic waste entering waterways reduces the amount of oxygen available and promotes the growth of harmful organisms (Bhada-Tata and Hoornweg 2016). Marine pollution is also increasing as a result of mismanaged solid waste on land, poor disposal practices by sea vessels, and runoff from sewage and polluted streams. Universal plastic usage is also leading to increasing nonbiodegradable waste litter in natural environments. (Please refer to case study 16 in chapter 7 for more information on marine litter and box 6.1 for information on plastic waste.)

A study focused on Southeast Asia estimated the economic cost of uncollected household waste that is burned, dumped, or discharged to waterways to be US$375/tonne (McKinsey 2016). For the same region, the World Bank estimated the integrated waste management costs for basic systems meeting good international hygienic standards to be US$50–US$100/tonne.

An environmental assessment can help governments understand the costs of solid waste management and its impacts on the environment as well as potential downstream issues. Of the cities studied with available data, 29 reported having completed an environmental assessment in the past five years, while 73 cities reported that no formal environmental assessment had been conducted in the past five years.

Box 6.1   Plastic Waste Management

In 2016, the world generated 242 million tonnes of plastic waste—12 percent of all municipal solid waste. This waste primarily originated from three regions—57 million tonnes from East Asia and the Pacific, 45 million tonnes from Europe and Central Asia, and 35 million tonnes from North America.

The visibility of plastic waste is increasing because of its accumulation in recent decades and its negative impact on the surrounding environment and human health. Unlike organic waste, plastic can take hundreds to thousands of years to decompose in nature (New Hampshire Department of Environmental Services n.d.). Plastic waste is causing floods by clogging drains, causing respiratory issues when burned, shortening animal lifespans when consumed, and contaminating water bodies when dumped into canals and oceans (Baconguis 2018). In oceans, plastic is accumulating in swirling gyres that are miles wide (*National Geographic* n.d.). Under ultraviolet light from the sun, plastic is degrading into "microplastics" that are almost impossible to recover and that are disrupting food chains and degrading natural habitats (United States NOAA n.d.). The Ellen MacArthur Foundation (2016) anticipates that, by weight, there will be more plastic in the oceans than fish by 2050 if nothing is done.

Plastic waste mainly enters the environment when it is poorly managed, such as through open dumping, open burning, and disposal in waterways. Unfortunately, with more than one-fourth of waste dumped openly and many formal disposal sites managed improperly, plastic litter is increasing. Even when plastic waste is collected, many countries lack capacity to process the waste. In 2017, Europe exported one-sixth of its plastic waste, largely to Asia (*The Economist* 2018).

There are many ways to curb plastic waste—by producing less, consuming less, and better managing the waste that already exists to prevent contamination or leakage. Taking these actions requires engagement from numerous stakeholders in society, including citizens, governments, community organizations, businesses, and manufacturers. Policy solutions, increased awareness, and improved design and disposal processes, among others, can minimize the impact of plastic waste on society.

Policy: Before pursuing dedicated plastics management solutions, governments must first focus on holistic management of waste. Cities need consistent collection services, safe and environmentally sound disposal, and consistent enforcement of policy before targeted interventions for plastic can be fully effective. Without strong basic waste management systems, plastic is likely to continue to be dumped when uncollected, citizens and businesses are less likely to comply with restrictions on materials for consumption or manufacturing, and cost recovery for waste systems will continue to be a struggle. With adequate primary waste management services in place, many cities have succeeded in focused interventions. For example, San Francisco, United States, implemented a plastic bag ban that led to a 72 percent decrease in plastic litter on local beaches from 2010 to 2017 (*Mercury News* 2018). In Rwanda and Kenya, plastic bag bans have been implemented effectively with financial and other legal penalties (de Freytas-Tamura 2017). In 2018, the European Union launched a strategy called Plastic Waste that aims to make all plastic packaging recyclable by 2030 and to ensure that waste generated on ships is returned to land (EU 2018b). However, innovative policies concerning plastic will not solve the issue of plastic mismanagement without proper institutions, systems, and incentives.

Society: Management of plastic waste often starts at the household and individual levels, and strategies to educate and motivate citizens can dramatically change behavior. In Jamaica, community members that serve as Environmental Wardens sensitize their neighbors about local cleanliness and safe and environmentally friendly disposal of waste. Environmental Wardens are community

(Box continues on next page)

Box 6.1   **Plastic Waste Management** *(continued)*

members employed by the Jamaican National Solid Waste Management Authority through a World Bank–supported project (Monteiro and Kaza 2016). Their role is to spread awareness about waste management and to keep communities clean and healthy. The communities and schools that are part of the project collect plastic bottles in large volumes, through competitions, and remove plastic litter from shared spaces and drains. They sell the collected plastic bottles to recyclers.

Industries: Plastic waste can be reduced or put to productive use at both a local and a global scale. Industries can alter manufacturing processes to reduce the amount of material needed, use recycled materials as inputs, or design new materials that can be degraded or more easily recycled. At a local level, recovered plastic can be used as inputs to make cement blocks, roads, and household goods such as baskets and mats (Growth Revolution Magazine 2009). These outlets for productive use can, in turn, drive increased collection and recovery of plastic waste. With about half of the plastic ever manufactured having been produced in the past 15 years, the collaboration of industry in reducing production and improving recycling is increasing in importance (*National Geographic* 2018).

Climate Change Mitigation

One of the major ways that solid waste contributes to climate change is its generation of greenhouse gas (GHG) emissions. The 1.6 billion tonnes of carbon dioxide–equivalent (CO_2-equivalent) emissions estimated for 2016 are anticipated to increase to 2.6 billion tonnes by 2050. Emissions from solid waste treatment and disposal, primarily driven by disposal in open dumps and landfills without landfill gas collection systems, were calculated using the CURB tool,[1] and they account for about 5 percent[2] of total global GHG emissions (World Bank 2018a; Hausfather 2017). GHG emissions result from inadequate waste collection, uncontrolled dumping, and burning of waste. Waste releases methane gas when disposed of in an oxygen-limited environment such as a dump or a landfill and releases pollutants and particulate matter during inefficient transportation and burning. Methane, generated from decomposing organic waste, is the solid waste sector's largest contributor to GHG emissions. It is many times more potent than CO_2.[3] Efforts to formalize the management of waste can significantly reduce GHG emissions. For example, a study by Zero Waste Europe concluded that the European Union could eliminate as much as 200 million tonnes of GHG emissions per year by 2030 with improved waste management practices (Ballinger and Hogg 2015).

Progress has been made in recent years. According to a United Nations Framework Convention on Climate Change report, from 1990 to 2015 the waste sector experienced the largest relative decrease in GHG emissions, at 20 percent, compared with other sectors (UNFCCC 2017). The emissions decrease is in part attributed to the growing effort of many cities to undertake mitigation activities in solid waste management. The Carbon Disclosure

Project shows that nearly 50 cities across the world have adopted mitigation measures in their climate plans (Carbon Disclosure Project 2013; IPCC 2007a). Looking ahead, more than 80 countries have identified solid waste management as an intervention area in their Nationally Determined Contributions, which are global commitments made by each country to mitigate and adapt to climate change under the historic United Nations Framework Convention on Climate Change agreement (Kampala Waste Management 2017).

Emissions can be mitigated through improved waste collection, waste reduction, reuse of products, recycling, organics waste management, and capture of GHGs for flaring or energy recovery. Reducing collection fleet lag times, improving routing efficiency such as through the use of geographic information systems, selecting cleaner fuels, and using fuel-efficient vehicles are potential approaches to reducing transportation emissions (Seto et al. 2014). Composting and anaerobic digestion are organic waste treatment options that prevent the generation of methane or its release into the atmosphere. Where landfills are used, the associated methane gas can be captured and flared, converted to power, used to heat buildings, or utilized to serve as fuel for vehicles. Waste-to-energy incinerators, which are relatively more complex and expensive, can reduce GHG emissions while generating electricity or thermal energy when operated effectively and to environmentally sound standards. A World Bank study in Indonesia shows that even basic improvements, such as increasing waste collection rates to 85 percent from 65 percent and introducing controlled landfilling for waste disposal, reduces GHG emissions by 21 percent (World Bank 2018b). These GHG reductions from the waste sector are an important element of Indonesia's committed Nationally Determined Contributions to the Paris Climate Agreement (Government of Indonesia 2016).

Climate Change Resilience

In the long term, the global community should consider solid waste resilience in addition to mitigation. As climate patterns change, waste management systems must prepare for extreme weather patterns that may cause waste to clog drainage systems during floods, landfills and dumps to collapse under heavy rains, or damage to urban infrastructure that may dramatically increase waste volumes. Cities should aim to ensure that their collection, transportation, and disposal systems can function regardless of the shock they face and should site the facilities to be resilient.

Climate change resilience at the local level may include the following planning and policy actions:

- Careful site selection for waste disposal based on topography and geology, natural resources, sociocultural factors, natural disaster patterns, and economy and safety (Al-Jarrah and Abu-Qdais 2006). For instance, a risk assessment can be done in flood-prone cities to determine the location, design, construction, operation, and decommissioning of waste facilities (Winne et al. 2012).

- Sufficient waste management capacity to meet projections of the city's current and anticipated growth.
- Emergency disposal sites and stakeholders for disposal of excessive amounts of waste to ensure systems function in times of disaster.
- Identification of vulnerabilities in existing infrastructure to prevent failure of facilities, and necessary investments in maintenance and upgrades.
- Formal education, community awareness efforts, and government incentives to promote responsible waste disposal and reduction to prevent litter that may clog drains or affect surrounding areas.

Circular Economy

Efforts to move toward a circular economy are gaining momentum, particularly in Europe. The circular economy model aims to use waste streams as a source of secondary resources and to recover waste for reuse and recycling. This approach is expected to achieve efficient economic growth while minimizing environmental impacts (Halkos and Petrou 2016).

In a circular economy, products are designed and optimized for a cycle of disassembly and reuse. The intention is to extend the lifespan of consumables and to minimize the environmental impact of final disposal. For challenging products, such as computers that are subject to rapid technological advancement and other durables containing metals and plastics that do not easily degrade, better disposal solutions and reuse could be part of the design process from the start.

In December 2015, the European Commission adopted a European Union Action Plan for a circular economy (EU 2015). In 2018 the European Union adopted a set of measures that support the implementation of the Action Plan and the European Union's vision of a circular economy (EU 2018a). These measures do the following:

- Set a goal to make all plastic packaging recyclable by 2030 and describe a holistic strategy to improve the economics and quality of plastics recycling
- Present ways to integrate legislation on waste, consumer products, and chemicals
- Outline 10 key indicators for monitoring progress in moving toward a circular economy across production, consumption, waste management, and investments
- Describe actions for more circular consumption of 27 of the most common materials used in the economy

Outside of Europe, the concept of a circular economy is slowly being embraced by national and local governments, and sometimes drives the development of goals and investments.

Technology Trends

As technology changes the way people live, communicate, and transact, it also affects the way waste is managed around the world. Governments and companies that manage waste integrate technologies at all steps of the value chain to reduce costs, increase materials for energy recovery, and connect with citizens.

Despite the ability of technological solutions to improve the way resources are used and recycled, technology selection differs by context. Communities vary by geography, technical capacity, waste composition, and income level and often the best solution is neither the newest nor the most advanced technologically. Whereas a mobile app may be the simplest way to inform citizens on service changes in an affluent city, technologies such as radio advertisements may be optimal in neighborhoods with high illiteracy rates. The following section reviews several of the simple and advanced technologies that have emerged to improve waste management around the world.

Data Management

Data are increasingly serving as the basis for decision making in waste management. From information on the layout and characteristics of local neighborhoods and the activity of collection trucks to data on recovery of waste fees, accurate information allows governments and operators to design and run more efficient operations and save money.

Formal information systems are increasing in cities but are not universal. From the data collected for this report, 29 percent of countries reported the existence of an information system. At an urban level, 49 cities reported an established information system, 89 did not have one, and 231 countries lacked data on information systems.

As their capacity has increased, many agencies have developed or improved central information systems to improve planning and to transparently monitor performance. For example, in Quito, Ecuador, La Empresa Metropolitana de Aseo (the Metropolitan Cleaning Company) has developed a central data management system that tracks collection routes, generates reports on service performance metrics, and allows citizens to report infractions of waste regulations (Sagasti Rhor 2016). In Japan, a central data system connects waste facilities around the country to a central national waste information system (Kajihara 2017). Measurements of toxins and emissions are reported in real time to the central database. Any problems in equipment operations trigger automatic reports to the plant operator so that emergencies can be addressed immediately. Other typical uses of information systems are detailed in box 6.2.

When it comes to data, solutions need not be complex. Simple data tools such as the Excel-based Data Collection Tool for Urban Solid Waste Management, developed by the World Bank and the Climate and Clean Air Coalition, can guide cities and planners in understanding the local waste situation quantitatively and comprehensively (World Bank 2013).

Box 6.2   **Examples of Information That Can Be Aggregated Using a Waste Management Data System**

- Real-time locations and routes of collection vehicles
- Weight of waste disposed of at different locations
- Emissions of landfills or waste-to-energy facilities
- Records of user payments
- History of waste collection at households
- Video streams of activities of waste equipment
- Radio and email communications with staff
- Registration of waste pickers
- Feedback from citizens
- Inventory of facilities and equipment

Waste Reduction and Manufacturing

Technology has been commonly used to support reduction of the amount of waste generated globally and to change manufacturing processes to reduce waste or to increase recyclability. In the Republic of Korea, radio-frequency identification (RFID) chips are often used to motivate citizens to reduce the waste that they generate. These small radio chips are embedded in personal cards that citizens use to open dumpsters and log the weight of the waste that they dispose of. Citizens are billed by the weight registered on the chip and are motivated to reduce the waste that they produce as a result. Korea's RFID-focused approach and overarching information management system are detailed in case study 17 in chapter 7.

Technologies used in manufacturing aim to enable the reuse of materials or decrease the use of virgin materials. Packaging innovations such as bio-degradable forks and bags reduce plastic waste and sometimes allow users to compost these materials. However, new materials require appropriate management, and poorly managed biodegradable packaging can lead to increased GHG emissions when landfilled or dumped and can fail to degrade fully in the wrong conditions (Vaughan 2016). Software is available that allows manufacturers to take waste into consideration in the product design process and to choose materials that have the least impact on the environment (Building Ecology n.d.). Some companies have developed processes that use waste materials as inputs for other products, such as using plastic and textiles to create new garments. Finally, new platforms are emerging that create a marketplace for used goods, thus reducing the need to manufacture new products (Sustainable Brands 2017).

Waste Collection

Waste collection and associated transportation is often the costliest step in waste management, and technology is extensively available to increase efficiency. Starting with the use of a geographic information system, a city can optimize routing and minimize improper use of trucks (Longhi et al. 2012).

Photo 6.1   **Solar-Powered Waste Compaction Bins in the Czech Republic**

Sensors can optimize routes and reduce unnecessary pickups. Dumpster sensors can signal how full a dumpster is so that pickups can be made accordingly. Solar-powered compactor bins use solar power to compact waste to one-sixth of its original volume and can alert the municipality or waste collector when a sensor detects that the bin is reaching capacity.

In locations with sufficient infrastructure, relevant geographical conditions, and ample financing, alternative approaches are being used for waste collection. Automation for waste collection vehicles ranges from the lifting of bins placed in the back of the truck to mechanical side arms that automatically pick up standardized bins directly from households. Even more automated collection solutions are being tested. In a limited number of areas with restricted transportation access or that are extremely dense, a more unusual approach could be pneumatic waste collection. In Roosevelt Island, New York City, United States, in an attempt to establish a car-free island, pneumatic waste collection was set up underground so that residents in high-rise buildings could place waste in a chute in their buildings that would be sucked into a tube, via a vacuum, to a central point for treatment and disposal (Chaban 2015). Although a pneumonic system could be a healthier and less congested alternative to truck-based collection systems, hurdles to adopting pneumonic systems could include cost and the inability to install needed infrastructure given a city's existing layout and substructure.

Mobile applications are also being implemented to assist in urban waste collection systems. Mobile applications are being used to inform citizens of collection schedules, source-separation guidelines, and fees. I Got Garbage is an example of a mobile application in India that is used by households to request waste collection services (box 6.3).

Box 6.3   I Got Garbage

I Got Garbage is an organization operating in Indian cities that uses an online platform to match waste pickers with households and businesses seeking waste services. The organization has successfully created and equipped waste social enterprises with the necessary skills for impact at scale. I Got Garbage supports more than 10,000 waste pickers and offers waste services from waste collection to local organics management and value-added recycling.

Figure B6.3.1   Features of I Got Garbage Application

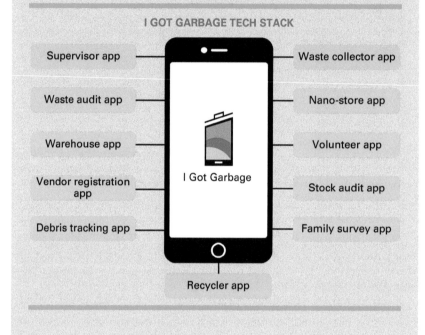

I GOT GARBAGE TECH STACK

Supervisor app — Waste collector app
Waste audit app — Nano-store app
Warehouse app — Volunteer app
Vendor registration app — Stock audit app
Debris tracking app — Family survey app
I Got Garbage
Recycler app

Waste Treatment and Disposal

Technology is being used in a variety of ways to improve waste treatment and disposal. However, the range of optimal technologies varies greatly by income level and local characteristics. More detailed information and guidance regarding solid waste management treatment and disposal technologies can be found in the World Bank's *Decision Maker's Guides for Solid Waste Management Technologies* (Kaza and Bhada-Tata 2018).

Low-Income Countries

Although open dumping and burning are common in low-income countries, there is a growing trend toward improving recycling and disposing of waste in controlled or sanitary landfills. Recycling is typically done by the informal sector in an unorganized fashion. Small-scale or even household biogas systems are also increasing in prevalence.

Middle-Income Countries

Landfills are the most common final disposal method in middle-income countries and are generally anticipated to continue being so. Improvements in recycling and in organics management are increasing. For recycling, sorting plants that involve manual or some form of automated sorting are becoming more common.

Primarily because of problems with the availability of land, large metropolitan cities in middle-income countries are looking at ways to avoid the development of large sanitary landfill sites, often far away from the city center, and to develop waste-to-energy incineration schemes instead. High land prices and often-elevated levels of electricity feed-in tariffs can be an important push for these investments. However, high costs, usually significantly above current cost levels, and the high organic composition of the waste, meaning that it is low in calorific value, also could present challenges to implementation. However, it is expected that modern waste incinerators could be built in some middle-income countries in the coming years. In China, quick development of incineration capabilities has already occurred, and the practice has become increasingly common in larger cities in Eastern China (Li et al. 2016).

High-Income Countries

Sanitary landfills and incinerators are prominent in high-income countries. High-income countries experience greater recovery and reintegration of materials from recycling and organics and use of byproducts such as refuse-derived fuel or other energy from waste than lower-income countries. Waste-derived energy is used for a range of purposes, such as in industry or to power waste facilities or buses. Automated landfill monitoring has increased, with some sites even using drones to assess the capacity of cells (Lucero et al. 2015).

High-income countries are making a substantial effort to recover materials from the source, with an emphasis on recycling and productive use of organic waste. Automation in recycling centers ranges from a conveyor belt to use of optical lasers and magnetic forces to separate waste (Peak 2013). Citizen participation for source separation of waste is common for smaller communities of less than 50,000 inhabitants, and mechanical sorting is commonly used for large cities. Greater attention is also being placed on management of food and green waste, sometimes through windrow composting, in-vessel composting, anaerobic digestion, and waste-to-liquid technologies. These technologies allow organic waste to be used effectively through capture of biogas and creation of a soil amendment or liquid fertilizer. These advances are complemented by improvements to distributed waste management, which emphasizes household interventions such as source separation.

Some solutions are less well-known or are still being piloted. A bioreactor landfill is a type of sanitary landfill that involves recirculation of leachate to more quickly degrade organic waste than in natural situations, increase landfill gas generation in a concentrated period, and reduce final leachate treatment, under certain conditions (Di Addario and Ruggeri

2016). Nonlandfill solutions that have been available for some time but have not been applied at large scale with municipal solid waste include advanced thermal technologies such as pyrolysis, gasification, and plasma arc technologies (Rajasekhar et al. 2015). These thermal processes break down waste with high temperatures in a zero- or low-oxygen environment with one of the main outputs being a synthetic gas. When these processes are applied to municipal solid waste, commercial and technical viability has shown mixed results, with multiple failed attempts.

A number of countries, particularly Japan, Korea, and some countries in Western Europe, have almost completely moved away from landfilling, and aim to reduce incineration and maximize waste reduction and recycling. With recycling reaching 50 percent and more in a few Western European countries, trading of household waste across countries for incineration is increasing.

Citizen Engagement

The success of sustained solid waste management is critically linked with public engagement and trust. Waste managers rely on citizens to consciously reduce the amount of waste they generate, separate or manage specific waste types at home, dispose of waste properly, pay for waste management services, and approve new disposal sites. To motivate this support, governments must gain the trust of citizens. Cities and countries are engaging the public by providing high-quality services that earn approval and trust and that, in turn, motivate citizens to pay for services, be environmentally aware, and comply with guidelines and regulations. Although changing citizen behavior can take time, the benefits of a strong relationship with the public are invaluable to a waste management system.

Education

Educational programs are a key aspect of raising awareness for solid waste. Many countries reach citizens using media. Effective programs distribute content in a variety of languages and through both basic and advanced technology, such as radio, television, and mobile phone applications. Other governments focus on schools to educate young citizens who will eventually become environmentally conscious adults. For example, in Kingston, Jamaica, school programs incorporate environmental and waste management issues into the formal curriculum and participate in hands-on activities such as onsite recycling, composting, and gardening. Vegetables grown in school gardens are used at the schools or given to students (Clarke, personal communication 2017). Some schools also encourage students and citizens to visit waste facilities such as recycling centers or landfills.

Of the countries and cities studied, several make waste management information available to the public. The most common types of information made available include collection schedules and waste drop-off locations, budgets and fees, local statistics on waste generation and composition, and community programs and recycling campaigns. For example, Bangkok, Thailand, publishes the Bangkok State of the Environment Report periodically, providing a comprehensive review of solid waste management in the city (Bangkok Metropolitan Administration 2012). Yokohama, Japan, reports on GHG emissions resulting from waste; Bern, Switzerland, provides recycling information specifically for visitors and migrants; and Montevideo, Uruguay, provides guidance to households on how to request a waste bin (City of Yokohama n.d.; Hello Switzerland n.d.; City of Montevideo n.d.).

Countries typically share information on national waste management statistics, legislation and policies, fees, and infrastructure such as landfills and transfer stations. Common platforms for information distribution include face-to-face interactions, signage, media, websites, periodic reports, mobile applications, physical booklets, and fliers. The city of Baltimore, United States, has even installed a trash collection machine with a humorous appearance in the city's harbor to catch the attention of citizens (box 6.4).

Box 6.4   Mr. Trash Wheel

Mr. Trash Wheel is a trash interceptor in Baltimore, Maryland, United States, that picks up litter floating in the Inner Harbor of Baltimore (Waterfront Partnership of Baltimore n.d.). Its remarkable visual appearance builds public awareness of proper waste management. The instrument's rotor is powered by water and solar energy, and it deposits floating waste into a dumpster behind the vessel using a moving conveyer belt.

Source: Photo courtesy of Waterfront Partnership of Baltimore; additional permission required for reuse.

Citizen Feedback

Governments benefit when citizens provide feedback on waste management services. Citizen feedback allows waste management agencies to measure satisfaction and trust, understand gaps in services, and make critical changes that benefit the population, the environment, and the economy.

For example, in Morocco, five cities launched a Citizen Report Card program, covering 25 percent of Morocco's urban population, to understand citizen satisfaction with waste operations. The results of the survey are used to evaluate whether private operators are performing well and to make decisions on renewing their contracts. In Maputo, Mozambique, MOPA is a digital platform accessible via phone, smartphone, and computer that allows citizens to report issues such as overflowing dumpsites (Vasdev and Barroca 2016). Citizens that provide feedback are notified once the issue is resolved. These forms of citizen engagement allow for a closed loop between public agencies and the community affected by services.

The cities studied use a variety of channels to collect citizen feedback, including phone, website, email, social media, surveys, and physical handouts. Toronto's online platform is detailed in case study 11 in chapter 7.

Financial Incentives

Financial incentives are a powerful tool for motivating sustained behavior change. Governments and organizations have used various mechanisms to tie financial incentives to participation in the waste management system. Financial incentives can be linked to source separation, waste collection, reduction in the volume of waste disposed of, and disposal according to designated locations and schedules.

For example, in Ningbo, China, results-based financing is being used to encourage households in high-rise apartment complexes to separate organics and recyclables (Lee et al. 2014). The government saves money since less waste is landfilled and returns a portion of the savings as a financial incentive for citizens who separate their waste. In addition, some cities only charge citizens for the disposal of residual waste or set fees for mixed waste disposal that are higher than fees for recycling services. In Kitakyushu, Japan, the government provides compost bins to households and holds public composting seminars that thousands of citizens have attended. Managing organic waste at the household level is cheaper for Kitakyushu citizens than paying by volume for formal disposal services (Matsuo n.d.).

Several organizations and companies have adopted the concept of personal rewards to encourage environmental engagement and change public behavior. There are websites where citizens can earn points for taking environmentally friendly actions, such as recycling or participating in a learning program, and then can use their points to earn discounts at stores or make donations to community organizations (Recyclebank, n.d.).

Social Impacts of Waste Management and the Informal Sector

The quality of solid waste management affects the urban poor in critical ways, with impacts on their health, housing quality, service access, and livelihoods. In urban low-income neighborhoods, up to two-thirds of solid waste is not collected (Baker 2012). In areas with poor service coverage, the incidence of diarrhea is twice as high and acute respiratory infections are six times higher than in areas with frequent waste collection (UN-Habitat 2010). Waste is often dumped or burned, releasing toxic airborne chemicals and liquid runoff that contaminates water sources (Akinbile and Yusoff 2011). The dumped waste can also be a source of food and shelter for rats, mosquitoes, and scavenging animals, which could carry diseases such as dengue fever. The homes closest to dumpsites are often those of vulnerable populations who make a living by scavenging for recyclables with a monetary value. Just as gaps in solid waste services disproportionately affect the poor, improvements in service delivery can dramatically improve the lives of vulnerable populations.

Informal Sector in Solid Waste Management

Informal waste recycling is a common livelihood for the urban poor in low- and middle-income countries. About 1 percent of the urban population, or more than 15 million people, earn their living informally in the waste sector (Medina 2010). In urban centers in China alone, about 3.3 million to 5.6 million people are involved in informal recycling (Linzner and Salhofer 2014).

Photo 6.2   Informal Recyclers in the Middle East and North Africa Region

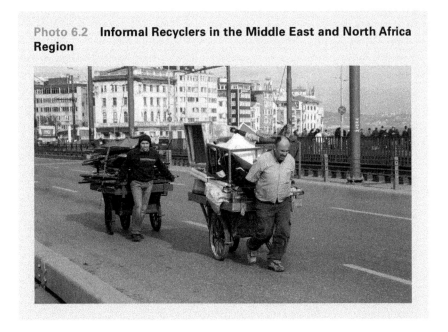

Waste pickers are often a vulnerable demographic and are typically women, children, the elderly, the unemployed, or migrants. They generally work in unhealthy conditions, lack social security or health insurance, are subject to fluctuations in the price of recyclable materials, lack educational and training opportunities, and face strong social stigma.

The data collected for this report revealed that in many places, the number of female waste pickers outnumbered the number of male waste pickers. For example, in Vientiane, Lao PDR, and Cusco, Peru, 50 percent and 80 percent of waste pickers are female, respectively (Keohanam 2017; Arenas Lizana 2012). Furthermore, many waste pickers are children who face greater risks to physical development and loss of education than adults. In Gjilan, Kosovo, about 40 percent of waste pickers at the local dumpsite are children (Kienast-Duyar, Korf, and Larsen 2017).

When properly supported and organized, informal recycling can create employment, improve local industrial competitiveness, reduce poverty, and reduce municipal spending on solid waste management and social services (Medina 2007). UN-Habitat found that waste pickers commonly collect 50–100 percent of waste in cities in low-income countries, at no cost to municipalities (UN-Habitat 2010). For example, waste pickers in Mumbai, India; Jakarta, Indonesia; and Buenos Aires, Argentina, are estimated to have an economic impact of more than US$880 million annually (Medina 2007). In Jakarta, waste pickers are estimated to divert 25 percent of the city's waste to productive use (Medina 2008). Some of the more successful interventions to improve waste pickers' livelihoods are formalization and integration of waste pickers, strengthening of the recycling value chain, and consideration of alternative employment opportunities (box 6.5).

Formalization and Integration of Waste Pickers

Formalizing informal waste pickers could lead to improved waste collection and recycling. Because of the social stigma often associated with waste picking, political buy-in could allow for social inclusion in the solid waste sector. National regulations or guidelines can lead to systematic consideration of waste pickers at all levels of government such as in Brazil (box 6.6). However, local municipalities are most directly empowered to provide recognition and social benefits to waste pickers, such as through legal identification, housing, health, and education.

Box 6.5   Waste Picker Cooperative Model: Recuperar

Members of Recuperar, a cooperative in Medellin, Colombia, earn 1.5 times the minimum wage. They can receive loans from the cooperative, are affiliated with the Colombian system of socialized medicine, have opportunities to earn scholarships to continue their studies, and are provided with life and accident insurance. The members mainly collect mixed waste and recyclables and, in 1998 alone, recovered 5,000 tonnes of recyclables (Medina 2005).

For instance, in Quezon City, the Philippines, approximately 3,000 waste pickers work at the Payatas landfill (Gupta 2014). They are provided with formal identification and work in shifts to allow each worker to earn income from the recovery of recyclables. Child labor is also banned. In Morocco, as part of a World Bank project, the government requires private sector solid waste management operations to employ any waste pickers that previously worked on the site. A clause is included in contracts to hold operators accountable (World Bank 2016).

One method governments can use to gain an understanding of the impact of the waste management system on multiple stakeholders is a social assessment (Bernstein n.d.). A social assessment analyzes several dimensions of the waste management system, from service quality to willingness to pay. Social assessments also address risks around informal labor, working conditions, and gender that are related to solid waste management. Insights from a social assessment can be used to improve the waste management system. In the study conducted for this book, 24 cities reported completion of a formal social assessment within the past five years, while 73 cities lacked a social assessment in the past five years. There is significant room for growth in cities' awareness and analytical assessment of the informal sector given existing challenges (box 6.7).

Box 6.6   Formalization of Waste Pickers in Brazil

Brazil passed a comprehensive Solid Waste National Policy in 2010, which both recognized waste picking cooperatives as service providers and created mechanisms to integrate informal waste workers into the country's formal system. The legislation's focus was on establishing safe disposal systems, decreasing waste generation, and increasing reuse and recycling, all through the combined efforts of the government, private, and informal waste sectors.

An overview of the legal framework in Brazil can be found at http://www.inclusivecities.org/wp-content/uploads/2012/07/Dias_WIEGO_PB6.pdf.

Box 6.7   Challenges for Waste Pickers

- In a study on waste pickers across five cities in Africa, Asia, and Latin America, 73 percent stated that lack of access to quality waste streams was a major or moderate problem (WIEGO 2014).
- About 87 percent stated that unstable prices were a major or moderate problem, and 61 percent found it difficult to negotiate better prices from buyers.
- In Bogota, Colombia, and Durban, South Africa, 80 percent of waste pickers said that harassment was a problem, 84 percent said they were treated poorly by local authorities, and 89 percent said that regulations and by-laws regarding waste are an issue. Some 97 percent identified social exclusion as a problem in their work.

Strengthening of the Recycling Value Chain

Waste pickers engage with the recycling value chain by collecting materials and selling them to middlemen who then clean and aggregate materials to distribute to industry. Governments and corporations can improve waste pickers' income prospects by creating sanctions to ensure fair prices from middlemen, directly offering waste pickers a fair and consistent wage at deposit centers, or helping waste picker cooperatives establish direct contracts with large buyers of recyclables, such as bottle manufacturers. An innovative partnership model between the private sector and waste pickers in Mexico is detailed in box 6.8.

Governments and nonprofits can also support waste pickers in forming organized cooperatives that provide a strong bargaining position with stakeholders (box 6.8). Micro and small enterprises and cooperatives help waste pickers increase the purchase price of their collected waste by negotiating with intermediaries and allow waste pickers to gain social recognition. There may also be opportunities to access infrastructure to provide additional value to the recyclables, such as baling or cleaning the materials.

Formal recognition also allows informal workers to gain job stability and acknowledgement of their work. Cooperative members consistently report a higher standard of living as well as improvements in self-esteem and self-reliance than when they work independently. In addition, organized workers are more productive and are healthier when provided with guaranteed collection routes and safe working conditions outside of dumpsites.

Box 6.8   Socially Responsible Plastics Recycling in Mexico

In 2006, the World Bank Group's International Finance Corporation (IFC) partnered with a Mexican company, PetStar, to finance a recycling plant that processes polyethylene terephthalate (PET), a common material in plastic bottles and food packaging. Because of low recycling rates in Mexico, sourcing a steady supply of raw PET from municipal recycling programs was not feasible. As a result, the project identified waste pickers as natural partners in collecting used PET products across the country. PetStar and the IFC worked together to generate socially responsible partnerships with waste picking communities that not only provided employment, but addressed issues with working conditions, organization and advocacy, and child labor. By partnering with the major beverage manufacturer Coca-Cola, PetStar found a guaranteed buyer for recycled plastic. This consistent revenue stream enabled PetStar to contract with informal workers at a fair, consistent wage. PetStar and the IFC's unique, vertically integrated approach to recycling is not only profitable but socially responsible for waste pickers and the environment.

Consideration of Alternative Employment Opportunities beyond Solid Waste Management

Integrating waste pickers into the solid waste management system might not always be efficient or even preferred by waste pickers. If local recycling markets are weak or if the waste collection or sorting needs of the city do not require extensive labor, waste pickers might be more productively employed outside of the waste management system. Since waste pickers often lack skills for alternative livelihoods, external employment requires social support and vocational training to ensure a smooth transition.

Job retraining or skill-building programs, in combination with social support programs such as in health care and child education, can support adult career transitions and minimize periods of vulnerability. Although the personalized attention and resources needed to support alternative livelihoods can be substantial, when provided properly, this support can help break the cycle of poverty for several future generations. An example of education reducing waste picking is that of the conditional cash transfer program, Bolsa Familia, in Brazil (Dias 2008; Medina 2007). It entailed giving a financial incentive to vulnerable families for sending their children to school and resulted in more than 40,000 children leaving waste picking to attend school.

Notes

1. CURB: Climate Action for Urban Sustainability is a low-carbon planning tool available at worldbank.org/curb.
2. Emissions estimated exclude waste-related transportation.
3. Methane has much higher short-term global warming potential (GWP) than CO_2. Over the typically used 100-year time horizon, methane has 25 times higher GWP, but over the shorter time frame of 20 years, methane has 72 times higher GWP than CO_2 (IPCC 2007b).

References

Akinbile, Christopher O., and Mohd S. Yusoff. 2011. "Environmental Impact of Leachate Pollution on Groundwater Supplies in Akure, Nigeria." *International Journal of Environmental Science and Development* 2 (1): 81–86. http://www.ijesd.org/papers/101-F10106.pdf.

Al-Jarrah, O., and H. Abu-Qdais. 2006. "Municipal Solid Waste Landfill Siting Using Intelligent System." *Waste Management* 26 (3): 299–306. https://www.ncbi.nlm.nih.gov/pubmed/16019199.

Arenas Lizana, J. 2012. "Social Inclusion Plan for Waste Pickers in Cusco, Calca and Urabamba Provinces" ["Plan de Inclusión Social para Segregadores en las Provincias de Cusco, Calca y Urubamba"]. Regional Development Program [Programa de Desarrollo Regional (PRODER)].

Baconguis, Beau. 2018. "Stemming the Plastic Flood." A Break Free from Plastic Movement Report. https://www.breakfreefromplastic.org/wp -content/uploads/2018/04/Stemming-the-plastic-flood-report.pdf.

Baker, Judy. 2012. *Climate Change, Disaster Risk, and the Urban Poor: Cities Building Resilience for a Changing World.* Washington, DC: World Bank.

Ballinger, Ann, and Dominic Hogg. 2015. "The Potential Contribution of Waste Management to a Low Carbon Economy." Eunomia Research and Consulting, Bristol, UK. http://www.acrplus.org/images/publication /Contribution_low_carbon_economy/EN_Main-Report-.pdf.

Bangkok Metropolitan Administration. "Bangkok State of the Environment 2012 (revised edition)." Department of Environment. http://203.155.220 .174/pdf/BangkokStateOfEnvironment2012RevisedEdition.pdf.

Bernstein, Janice. n.d. "Social Assessment in Municipal Solid Waste Management." World Bank, Washington, DC. http://siteresources. worldbank.org/INTURBANDEVELOPMENT/Resources/336387 -1249073752263/6354451-1249073991564/bernsteinsa.pdf.

Bhada-Tata, Perinaz, and Daniel Hoornweg. 2016. "Solid Waste and Climate Change." *State of the World 2016: Can a City Be Sustainable?* Washington, DC: Worldwatch Institute. http://www.worldwatch.org /http%3A/%252Fwww.worldwatch.org/bookstore/publication/state -world-can-city-be-sustainable-2016.

Building Ecology. "Life Cycle Assessment Software, Tools and Databases." n.d. BuildingEcology.com. http://www.buildingecology.com/sustainability /life-cycle-assessment/life-cycle-assessment-software.

Carbon Disclosure Project. 2013. "CDP Cities 2013: Summary Report on 110 Global Cities." Carbon Disclosure Project, London. http://www.c40 .org/researches/c40-cdp-2013-summary-report.

Chaban. Matt. 2015. "Garbage Collection, without the Noise or the Smell." *New York Times*, August 3. https://www.nytimes.com/2015/08/04 /nyregion/garbage-collection-without-the-noise-or-the-smell.html.

City of Montevideo. n.d. http://www.montevideo.gub.uy/gestion-de-residuos.

City of Yokohama. n.d. http://www.city.yokohama.lg.jp/shigen/sub-data/.

de Freytas-Tamura, Kimiko. 2017. "Public Shaming and Even Prison for Plastic Bag Use in Rwanda." *New York Times*, October 28. https://www .nytimes.com/2017/10/28/world/africa/rwanda-plastic-bags-banned.html.

Di Addario, Martina, and Bernardo Ruggeri. 2016. "Fuzzy Approach to Predict Methane Production in Bioreactor Landfills." https://www .researchgate.net/publication/304334842_Fuzzy_Approach_to_Predict _Methane_Production_in_Bioreactor_Landfills.

Dias, Sonia. 2008. "Fórum Lixo e Cidadania: Catadores de Problema Social à Questão Sócio-Ambiental." Paper presented at the First World Conference of Waste Pickers, Bogotá, March.

The Economist. 2018. "The Known Unknowns of Plastic Pollution." March 3. https://www.economist.com/international/2018/03/03/the -known-unknowns-of-plastic-pollution.

Ellen MacArthur Foundation. 2016. *The New Plastics Economy— Rethinking the Future of Plastics.* Cowes, UK: Ellen MacArthur Foundation. https://www.ellenmacarthurfoundation.org/assets/down loads/EllenMacArthurFoundation_TheNewPlasticsEconomy_ Pages.pdf.

EU (European Union). 2015. "Closing the Loop—An EU Action Plan for the Circular Economy." European Commission, Brussels. https://eur-lex .europa.eu/legal-content/EN/TXT/?uri=CELEX%3A52015DC0614.

———. 2018a. "Implementation of the Circular Economy Action Plan." European Commission, Brussels. http://ec.europa.eu/environment/circular -economy/index_en.htm.

———. 2018b. "Plastic Waste: A European Strategy to Protect the Planet, Defend Our Citizens and Empower Our Industries." Press Release, January 16, Strasburg. http://europa.eu/rapid/press-release_IP -18-5_en.htm.

Government of Indonesia. 2016. "First Nationally Determined Contribution: Republic of Indonesia." http://www4.unfccc.int/ndcregistry/Published Documents/Indonesia%20First/First%20NDC%20Indonesia_submitted %20to%20UNFCCC%20Set_November%20%202016.pdf.

Growth Revolution Magazine. 2009. "Teresa: Waste Management." modelhttps://growthrevolutionmag.wordpress.com/2009/09/21 /teresawaste-management-model/.

Gupta, S. 2014. "Integrating the Informal Sector." *Handshake* 12: 66–71.

Halkos, George E., and Kleoniki N. Petrou. 2016. "Moving towards a Circular Economy: Rethinking Waste Management Practices." *Journal of Economic and Social Thought* 3 (2): 220–40. http://www.kspjournals .org/index.php/JEST/article/view/854/912.

Hausfather, Zeke. 2017. "Analysis: Global CO2 Emissions Set to Rise 2% in 2017 after Three-Year 'Plateau'." *CarbonBrief*, November 13. https:// www.carbonbrief.org/analysis-global-co2-emissions-set-to-rise-2-percent -in-2017-following-three-year-plateau.

Hello Switzerland. http://www.helloswitzerland.ch/-/waste-and-recycling-in -switzerland.

IPCC (Intergovernmental Panel on Climate Change). 2007a. "Climate Change 2007: Impacts, Adaptation and Vulnerability. Contribution of Working Group II to the Fourth Assessment Report of the Intergovernmental Panel on Climate Change." Edited by M. L. Parry, O. F. Canziani, J. P. Palutikof, P. J. van der Linden, and C. E. Hanson. Cambridge University Press, Cambridge, UK.

———. 2007b. "Climate Change 2007: The Physical Science Basis. Contribution of Working Group I to the Fourth Assessment Report of the Intergovern mental Panel on Climate Change." Edited by S. Solomon, D. Qin, M. Manning, Z. Chen, M. Marquis, K. B. Averyt, M. Tignor, and H. L. Miller. Cambridge University Press, Cambridge, UK, and New York. http://www .ipcc.ch/pdf/assessment-report/ar4/wg1/ar4_wg1_full_report.pdf.

Kajihara, Hiroyuki. 2017. "Introduction to the Municipal Waste Administration." Presented at the Solid Waste Management Technical Deep Dive in Tokyo, Japan, March.

Kampala Waste Management. 2017. "Kampala Waste Treatment & Disposal PPP Project." Investor Conference.

Kaza, Silpa, and Perinaz Bhada-Tata. 2018. "Decision Maker's Guides for Solid Waste Management Technologies." World Bank Urban Development Series, World Bank, Washington, DC.

Keohanam, B. 2017. Director, Urban Development Division, Department of Housing and Urban Planning, Ministry of Public Works and Transport, Government of Lao PDR. Personal communication with the World Bank, May 17.

Kienast-Duyar, U., N. Korf, and O. Larsen, eds. 2017. "Solid Waste Management in Kosovo: Assessment of a Waste Bank Model in Dardania, Pristina." Berlin University of Technology, German Corporation for International Cooperation [Deutsche Gesellschaft für Internationale Zusammenarbeit GmbH (GIZ)], University of Pristina, and Municipality of Pristina.

Lee, Marcus, Farouk Banna, Renee Ho, Perinaz Bhada-Tata, Silpa Kaza. 2014. *Results-Based Financing for Municipal Solid Waste.* Washington, DC: World Bank.

Li, Xinmei, Changming Zhang, Yize Li, and QiangZhi. 2016. "The Status of Municipal Solid Waste Incineration (MSWI) in China and Its Clean Development." *Energy Procedia* 104 (December): 498–503. https://www .sciencedirect.com/science/article/pii/S1876610216316423.

Linzner, R., and S. Salhofer. 2014. "Municipal Solid Waste Recycling and the Significance of Informal Sector in Urban China." *Waste Management and Research* 32 (9): 896–907. doi: http://dx.doi.org/10.1177/0734242 X14543555.

Longhi, S., Davide Marzioni, Emanuele Alidori, Gianluca Di Buo, Mario Prist, Massimo Grisostomi, and Matteo Pirro. 2012. "Solid Waste Management Architecture Using Wireless Sensor Network Technology." Conference Paper, 5th International Conference on New Technologies, Mobility and Security, May 7–10. https://www.researchgate.net /publication/261086497_Solid_Waste_Management_Architecture _Using_Wireless_Sensor_Network_Technology.

Lucero, Osvaldo, Maria E. Rey Nores, Ezequiel Verdini, and James Law. 2015. "Use of Drones on Landfills." ISWA Conference, Antwerp, Belgium, September 7–9. https://www.scsengineers.com/wp-content/uploads/2015 /10/Use_of_Drones_on_Landfills_092015_James_Law.pdf.

Matsuo, Yasushi. "Introduction of Food Waste Recycling from Business Activities in Kitakyushu City." Presented at World Bank Solid Waste Management Technical Deep Dive. Kitakyushu, Japan.

McKinsey. 2016. "The Circular Economy: Moving from Theory to Practice." McKinsey Center for Business and Environment Special Edition. https:// www.mckinsey.com/~/media/McKinsey/Business%20Functions /Sustainability%20and%20Resource%20Productivity/Our%20 Insights/The%20circular%20economy%20Moving%20from%20 theory%20to%20practice/The%20circular%20economy% 20Moving%20from%20theory%20to%20practice.ashx.

Medina, Martin. 2005. "Waste Picker Cooperatives in Developing Countries." El Colegio de la Frontera Norte, Mexico. http://www.wiego .org/sites/default/files/publications/files/Medina-wastepickers.pdf.

———. 2007. *The World's Scavengers: Salvaging for Sustainable Consumption and Production.* Lanham, MD: Altamira Press.

———. 2008. "The Informal Recycling Sector in Developing Countries: Organizing Waste Pickers to Enhance Their Impact." *Gridlines.* World Bank, Washington, DC. http://documents.worldbank.org/curated/en /227581468156575228/pdf/472210BRI0Box31ing1sectors01PUB LIC1.pdf.

———. 2010. "Scrap and Trade: Scavenging Myths." March 15, Our World, United Nations University, Tokyo. March 15. https://ourworld.unu.edu /en/scavenging-from-waste.

The Mercury News. 2018. "Success! California's First in the Nation Plastic Bag Ban Works." November 13. https://www.mercurynews.com/2017 /11/13/editorial-success-californias-first-in-the-nation-plastic-bag-ban -works/.

Monteiro, Emanuela, and Silpa Kaza. 2016. "Jamaica—Jamaica Integrated Community Development Project: P146460—Implementation Status Results Report: Sequence 05." World Bank, Washington, DC.

National Geographic. n.d. "Great Pacific Garbage Patch." *National Geographic,* Washington, DC. https://www.nationalgeographic.org /encyclopedia/great-pacific-garbage-patch/.

New Hampshire Department of Environmental Services. n.d. "Approximate Time It Takes for Garbage to Decompose in the Environment." New Hampshire DES, Concord, NH. https://www.des.nh.gov/organization /divisions/water/wmb/coastal/trash/documents/marine_debris.pdf.

Peak, Katie. 2013. "How It Works: Inside the Machine That Separates Your Recyclables." *Popular Science*, August 28. https://www.popsci.com/tech nology/article/2013-07/how-it-works-recycling-machines-separate -junk-type.

Rajasekhar, M., N. Venkat Rao, G. Chinna Rao, G. Priyadarshini, and N. Jeevan Kumar. 2015. "Energy Generation from Municipal Solid Waste by Innovative Technologies—Plasma Gasification." *Procedia Materials Science* 10: 513–18. https://www.sciencedirect.com/science /article/pii/S2211812815003326.

Recyclebank. n.d. https://www.recyclebank.com/.

Sagasti Rhor, Carlos. 2016. "Waste Collection IT Systems for Waste Management—Case: Quito—EMASEO." Presented at World Bank event "Citizen Engagement & ICT in Solid Waste Management," Washington, DC, February 18.

Seto, K. C., S. Dhakal, A. Bigio, H. Blanco, G. C. Delgado, D. Dewar, L. Huang, A. Inaba, A. Kansal, S. Lwasa, J. E. McMahon, D. B. Müller, J. Murakami, H. Nagendra, and A. Ramaswami. 2014. "Human Settlements, Infrastructure and Spatial Planning." In *Climate Change 2014: Mitigation of Climate Change: Contribution of Working Group III to the Fifth Assessment Report of the Intergovernmental Panel on Climate Change.* Edited by O. Edenhofer, R. Pichs-Madruga, Y. Sokona, E. Farahani, S. Kadner, K. Seyboth, A. Adler, I. Baum, S. Brunner, P. Eickemeier, B. Kriemann, J. Savolainen, S. Schlömer, C. von Stechow, T. Zwickel and J.C. Minx. Cambridge, UK, and New York: Cambridge University Press.

Sustainable Brands. 2017. "2018 Circulars Nominate 43 Changemakers Paving the Path to a Circular Future." http://www.sustainablebrands .com/news_and_views/next_economy/sustainable_brands/2018_circu- lars_nominate_43_change_makers_paving_path_.

Thompson, Andrea. 2014. "For Air Pollution, Trash Is a Burning Problem." Climate Central. http://www.climatecentral.org/news/where-trash-is-a -burning-problem-17973.

UNFCCC. 2017. "National Greenhouse Gas Inventory Data for the Period 1990–2015." United Nations Framework Convention on Climate Change. http://unfccc.int/resource/docs/2017/sbi/eng/18.pdf.

UN-Habitat. 2010. *Solid Waste Management in the World's Cities: Water and Sanitation in the World's Cities 2010.* London and Washington, DC: Earthscan. http://www.waste.nl/sites/waste.nl/files/product/files/swm_in _world_cities_2010.pdf.

United States NOAA (National Oceanic and Atmospheric Administration). n.d. "What Are Microplastics." United States NOAA, Washington, DC. https://oceanservice.noaa.gov/facts/microplastics.html.

Vasdev, Samhir, and Jean Paulo Gil Barroca. 2016. "Development into Practice: Co-Designing a Citizen Feedback Tool That Makes Sense." World Bank blog. https://blogs.worldbank.org/taxonomy/term/15794.

Vaughan, Adam. 2016. "Biodegradable Plastic 'False Solution' for Ocean Waste Problem." *The Guardian*, May 23. https://www.the guardian.com/environment/2016/may/23/biodegradable-plastic -false-solution-for-ocean-waste-problem.

Waterfront Partnership of Baltimore. n.d." Trash Wheel Project." http://balti morewaterfront.com/healthy-harbor/water-wheel/.

WIEGO. 2014. "The Urban Informal Workforce: Waste Pickers/Recyclers." http://www.wiego.org/sites/wiego.org/files/publications/files/IEMS -waste-picker-report.pdf.

Winne, S., L. Horrocks, N. Kent, K. Miller, C. Hoy, M. Benzie, and R. Power. 2012. "Increasing the Climate Resilience of Waste Infrastructure." Final Report under Defra contract ERG 1102. AEA Technology, Didcot, UK. https://www.gov.uk/government/uploads/system/uploads/attachment _data/file/183933/climate-resilience-full.pdf.

World Bank. 2013. Data Collection Tool for Urban Solid Waste Management. Excel Spreadsheet Tool. World Bank, Washington, DC.

———. 2016. "Morocco—Municipal Solid Waste Sector Development Policy Loans (3 and 4) Project (English)." World Bank, Washington, DC. http:// documents.worldbank.org/curated/en/563061482164575195/Morocco -Municipal-Solid-Waste-Sector-Development-Policy-Loans-3-and-4-Project.

———. 2018a. "The CURB Tool: Climate Action for Urban Sustainability." World Bank, Washington, DC. www.worldbank.org/curb.

———. 2018b. "Improvement of Solid Waste Management to Support Regional and Metropolitan Cities. Project Appraisal Document." World Bank, Washington, DC. https://operationsportalws.worldbank.org/Pages /WorkingDocuments.aspx?projectid=P157245.

Additional Resources

Babel, Sandhya, and Xaysackda Vilaysouk. 2015. "Greenhouse Gas Emissions from Municipal Solid Waste Management in Vientiane, Lao PDR." *Waste Management & Research* 34 (1): 30–37. http://journals .sagepub.com/doi/full/10.1177/0734242X15615425.

Camara, Jaime. 2016. "From Greenfield to World's Largest Food Grade PET Recycling Facility." Presented at a World Bank event "Finding Resilience in a Volatile Recycling Market."

Cohen, Peter, Jeroen Ijgosse, and Germán Sturzenegger. 2013. "Preparing Informal Recycler Inclusion Plans: An Operational Guide." Inter-American

Development Bank, Washington, DC. https://publications.iadb.org
/handle/11319/697?locale-attribute=en#sthash.AIIheLww.dpufhttp://
publications.iadb.org/handle/11319/697?locale-attribute=en.

Dias, Sonia. 2011. "Overview of the Legal Framework for Inclusion of
Informal Recyclers in Solid Waste Management in Brazil." WIEGO
Policy Brief (Urban Policies) No. 6. May. http://www.inclusivecities.org
/wp-content/uploads/2012/07/Dias_WIEGO_PB6.pdf.

I Got Garbage. https://www.igotgarbage.com/.

Inclusive Cities. n.d. "The Urban Informal Workforce: Waste Pickers
/Recycler." Informal Economy Monitoring Study. shttp://www.inclu
sivecities.org/wp-content/uploads/2012/08/IEMS-WP-Sector-Summary
-english.pdf.

IPCC (Intergovernmental Panel on Climate Change). 2014. "Human
Settlements, Infrastructure and Spatial Planning." In *Climate Change
2014: Mitigation of Climate Change. Contribution of Working Group III
to the Fifth Assessment Report of the Intergovernmental Panel on
Climate Change.* Cambridge University Press, Cambridge, UK, and
New York: IPCC.

McDonough, W., and M. Braungart. 2002. *Cradle to Cradle: Remaking the
Way We Make Things.* New York: North Point Press.

Schwarz, H. G., M. Meyer, C. J. Burbank, M. Kuby, C. Oster, J. Posey,
E. J. Russo, and A. Rypinski. 2014. "Transportation." In *Climate Change
Impacts in the United States: The Third National Climate Assessment,*
edited by J. M. Melillo, T. C. Richmond, and G. W. Yohe. Washington,
DC: U.S. Global Research Program.

Waste 360 http://www.waste360.com/route-optimization/how-techn
ology-continues-improve-waste-management.

World Economic Forum. 2014. *Towards the Circular Economy: Accelerating
the Scale-up across Global Supply Chains.* Geneva: World Economic
Forum. http://www3.weforum.org/docs/WEF_ENV_TowardsCircular
Economy_Report_2014.pdf.

World Bank. 2015. "Morocco Solid Waste Sector Development Policy Loan
4. Implementation Completion Report (ICR) Review." World Bank,
Washington, DC. http://documents.worldbank.org/curated/en/70073
1500904650697/pdf/ICRR-Disclosable-P148642-07-24-2017-15009
04641929.pdf.

———. 2016. "Waste Management Key to Regaining Public Trust in the
Arab World." World Bank, Washington, DC. http://www.worldbank.org
/en/news/feature/2016/03/14/waste-management-key-to-regaining
-public-trust-arab-world.

https://hubs.worldbank.org/docs/ImageBank/Pages/DocProfile
.aspx?nodeid=27051835.

CHAPTER 7

Case Studies

---

1. A Path to Zero Waste in San Francisco, United States

In 2002, San Francisco announced a vision to send zero waste to landfills by 2020. Through initiatives to promote recycling and composting, San Francisco is now one of the greenest cities in North America and a global leader in waste management (Economist Intelligence Unit 2011).

San Francisco's success has been achieved largely by robust public policy implemented by determined political leadership, strong public-private partnerships, resident education, and financial incentives for waste reduction.

San Francisco was the first city in the United States to implement strict legislation about the use of or management of specific materials. The city prohibited the use of styrofoam and polystyrene foam in food service (2006), required mandatory recycling for construction debris (2007), banned plastic bags in drugstores and supermarkets (2009), and implemented mandatory recycling and composting for both residents and businesses (2009). San Francisco most recently also banned the sale of plastic water bottles in 2014 (EPA 2017).

State-of-the-art outreach programs covering residences, businesses, schools, and events are widespread, and financial incentives encourage waste reduction and recycling. To help residents more clearly understand their waste disposal practices and financial impact, each house or building receives a detailed bill for waste management fees. Payments are reduced if residents shift their waste from mixed waste bins to ones designated for recycling or composting. Furthermore, the size of the provided mixed waste bins was halved and the size of recycling containers was doubled. Waste bins are regularly inspected, and households that fail to comply with policies first receive warnings, followed by a financial penalty.

San Francisco also introduced the first and largest urban food waste composting collection program in the United States, covering both the commercial and residential sectors. San Francisco has collected more than a million tons of food waste, yard trimmings, and other compostable materials and turned these materials into compost for local farmers and wineries.

As a result of its efforts, San Francisco achieved nearly 80 percent waste diversion in 2012—the highest rate of any major city in the United States (EPA 2017).

2. Achieving Financial Sustainability in Argentina and Colombia

A key challenge often faced by municipal solid waste management systems is a shortage of financial resources. This shortage is often caused by a lack of dedicated government funding, low fees that fail to fully cover costs, tariffs that are not enforced, and a shortage of data on the real cost of services. Argentina and Colombia are effectively achieving financial sustainability with the approaches discussed below:

Financial Sustainability in Argentinean Municipalities

As in many other Latin American countries, municipal governments in Argentina were not aware of real solid waste management costs because they did not have a standard methodology or accounting system for estimating them. Municipalities also generally did not charge fees for waste services and very little in the way of municipal funds was earmarked for solid waste management.

Argentina quantified the total cost of its waste system to improve long-term sustainability. Under the World Bank–financed Integrated Solid Waste Management Project, the Secretariat of Environment and Sustainable Development (SAyDS)[1] developed a tool known as the Integrated Urban Solid Waste Management Economic and Financial Matrix. This tool helps municipalities understand the real costs of services and value of investments. The tool analyzes each stage of the solid waste management value chain, identifies the proportion of costs recovered by fees, and identifies ways to reallocate budget resources to improve financial sustainability.

The tool was made available to all municipalities in Argentina. Based on its deployment, SAyDS and the Ministry of Environment set the following goals for municipalities:

- Calculation of all integrated solid waste management costs (figures 7.2.1–7.2.3) and identification of all associated revenues toward the goal of equilibrating waste management accounts
- Development of potential new cost-recovery schemes and calculation of the associated fees using data
- Implementation of the polluter-pays principle so that larger generators of waste pay more

Through in-person and online trainings, 535 municipal and provincial staff were trained and municipalities covering 26 percent of the population collected financial data using the tool.[2]

The municipalities of Mar del Plata, Rosario, Viedma, Concordia, and Posadas have implemented cost recovery systems using the financial matrix. Mar del Plata, a large coastal municipality, implemented a differentiated fee system across wealthy and poor neighborhoods after a broad communications campaign and outreach effort. Both the variable costs of the waste system and the operational costs of the landfill are covered. Rosario, on

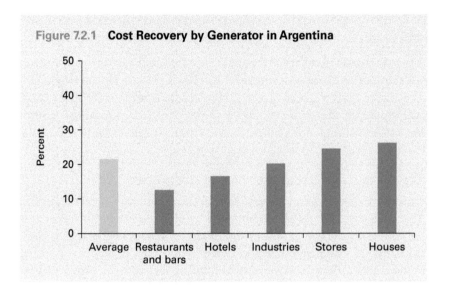

Figure 7.2.1   **Cost Recovery by Generator in Argentina**

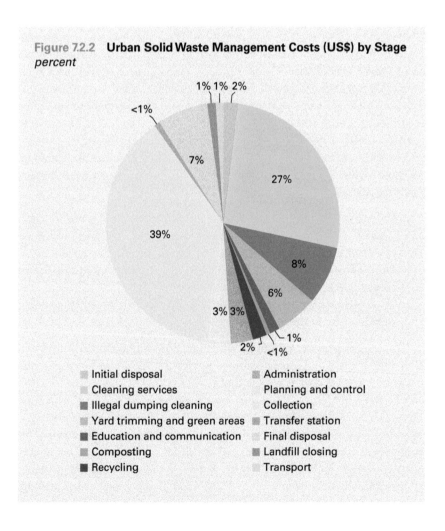

Figure 7.2.2   **Urban Solid Waste Management Costs (US$) by Stage**
percent

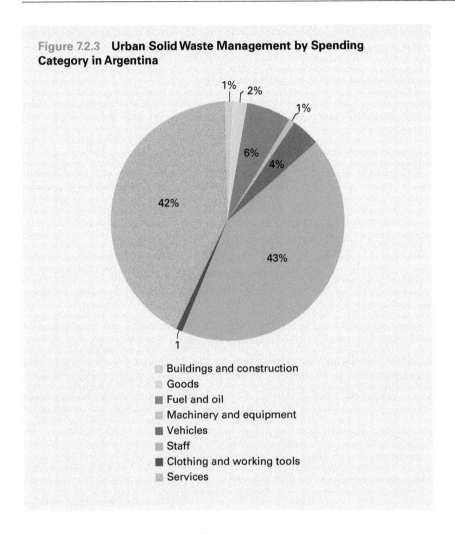

Figure 7.2.3 **Urban Solid Waste Management by Spending Category in Argentina**

1%   2%

1%

6%

4%

42%

43%

1

- Buildings and construction
- Goods
- Fuel and oil
- Machinery and equipment
- Vehicles
- Staff
- Clothing and working tools
- Services

the other hand, applied a specific fee to large waste generators. Municipal networks have also been developed to share information and experiences, such as suppliers that provide superior goods and services or that offer more competitive costs, peer-to-peer advice on strategy and operations, and opportunities for technicians from municipalities to participate in personnel exchanges with other towns and facilities within their province.

A key factor for implementation of the financial tool was having the necessary human and financial resources. SAyDS was fully staffed with qualified teams that could carry out outreach and capacity-building campaigns to provincial and municipal governments, tailor the training to specific needs of the local governments, and scale up the training nationwide. Through this success, municipalities built trust with the federal government and had the political support needed to improve cost recovery.

The tool was complemented by support for an institutional framework developed by the Integrated Solid Waste Management Project that allowed for agile coordination between the municipal, provincial, and federal governments.

Colombia's Strategies for Cost Recovery

In Colombia, the Regulatory Commission for Drinking Water and Basic Sanitation regulates public utilities involved in the distribution and processing of water and wastewater and also influences cost recovery for municipal solid waste systems. The commission established a national methodology for determining the maximum service fee that local service providers can charge to users. In 2016, the commission developed a formula that accounts for all costs in every step of the solid waste management system, including urban cleaning and sweeping, collection and transfer, final disposal, leachate management, and recycling (Correal 2016). This national framework enables municipalities to systematically recover their expenses and finance the services that they provide. In 2014, 84.5 percent of the sector's revenues originated from the collection of fees (Correal 2015).

An important success factor for this system is the involvement and authority of the central government. Through Article 370, the Colombian constitution assigns responsibility to the president for ensuring good administration and efficiency of public utilities through control, inspection, and surveillance. Law 142 allows municipal governments to recover the costs of local urban services. This legal infrastructure is complemented by the participation and involvement of stakeholders such as private contractors and recyclers.

According to the Ministry of Housing, City and Territory, by implementing tariff systems, 891 out of 1,122 Colombian municipalities, or nearly 80 percent, managed to recover costs from user fees by 2013 (Correal 2014). Colombia´s success in cost recovery through accounting, legal infrastructure, and institutional commitment can be replicated and adapted to other Latin American countries and regions around the world.

3. Automated Waste Collection in Israel

In 2012, a green neighborhood, Neot Rabin, was inaugurated in Israel's historic city of Yavne (Cohen 2012). Neot Rabin houses the country's first pneumatic waste collection system, which is also known as an automated vacuum collection (AVAC) system. Buildings with AVAC systems use a network of underground pipes to connect each residential unit with a centralized garbage storage unit. On each floor, residents dispose of waste in two garbage chutes: dry waste in one and wet waste in another. Garbage placed in these chutes is automatically directed to an underground storage unit.

Once a week, waste from residential buildings is pumped or vacuumed through a pipe at speeds of between 50 and 80 kilometers per hour to an aggregated storage center. The waste is stored in sealed containers in preparation for sorting and compaction. Finally, waste is transferred to containers that are removed by truck and transported to final disposal sites.

Based on the success of the pilot AVAC system in Neot Rabin, Yavne began replacing the municipality's public trash bins with pneumatic bins in 2014. As of 2015, about 30 pneumatic waste collection points were used in public areas, including parks, schools, and streets, providing immediate removal of waste from these spaces, eliminating waste-associated odor issues, and reducing traffic congestion. The municipalities of Ra'anana and Bat Yam also began assembling automated waste collection systems for residential buildings (Revolvy n.d.).

However, AVAC has certain limitations, such as the high initial investment required for establishing the system, operational difficulties when pipes are blocked, workforce training, public willingness to engage in separate disposal, and challenges to collection of bulky and electronic waste (Nakou, Benardos, and Kaliampakos 2014).

4. Cooperation between National and Local Governments for Municipal Waste Management in Japan

Japan manages its waste through comprehensive governance and advanced technologies. Of the nearly 44 million tonnes of waste generated annually, only 1 percent is landfilled. The remainder is recycled or converted to energy in state-of-the-art waste-to-energy facilities. Japan's efficient solid waste management practices can be largely attributed to effective cooperation between its national and local governments. The central and urban public authorities coordinate along several dimensions, from data collection to financing.

Data Collection and Database Management

Each year, the national Ministry of the Environment conducts an annual waste management survey. Local governments' responses are aggregated in a comprehensive database that both national and local governments use to develop plans, strategies, and policies. Information surveyed includes the quantity of waste that is generated and the amount of waste disposed of via recycling, composting, and incineration. The materials recovery rates reported through the survey are disclosed to the public, which provides incentives to local governments to increase sustainable disposal practices.

The transparent data system allows local governments to compare their plans and outcomes with those of other local governments that have similar economic and demographic profiles. Local governments use this

Table 7.4.1 Cooperation of National and Local Governments in Japan on Municipal Solid Waste Management

Task	Local governments	Relationship	National governments
1. Survey on the state of municipal solid waste management	Collection and submission of waste-related data	Waste Data → ← Database	Collect data from local governments and aggregate responses within a central database
2. Basic municipal waste management plan	Development of a solid waste management plan	Guidelines → ← Plan	Provision of guidelines for municipal solid waste management plans
3. Waste management plan implementation	Construction of waste treatment facilities	Construction → ← Subsidies	Provision of subsidies for construction of waste treatment facilities
4. Exchange of resources and information across government levels	Collect and submit feedback to national agencies	Information ←→ Human Resources	Facilitate exchange of human resources between national and local governments

Source: Shiko Hayashi.

information to evaluate and continually improve their processes. Members of the public and academic organizations may also use the data to evaluate the effectiveness of the waste management system.

In 2016, 1,741 municipalities and 578 special district authorities[3] completed the national survey.

Municipal Solid Waste Management Planning

All local governments in Japan are required to develop a local solid waste management plan that looks ahead about 10 years. To ensure consistency and thoroughness of local plans, the national government publishes guidelines for municipalities; these guidelines urge municipalities to detail their intended initiatives to sustainably treat waste and promote waste reduction, reuse, and recycling. All local governments comply with national laws and regulations, including the Air Pollution Control Act, the Soil Contamination Countermeasures Act, the Water Pollution Prevention Act, and the Act on Promotion of Private Finance Initiatives.

Financial Support for Municipal Solid Waste Infrastructure

The Japanese national government provides subsidies to municipalities to develop and improve waste treatment facilities based on the waste management plans submitted by local governments. Subsidies cover up to one-third of the cost of basic infrastructure projects, and for advanced facilities, such as high-efficiency waste-to-energy facilities, subsidies often cover half of project costs. The types of projects that are subsidized include recycling facilities, waste-to-energy plants, organic waste processing sites, septic tanks, landfills, refurbishing of waste treatment equipment, and extension of the lifespan of existing waste disposal facilities.

The remaining capital costs are the responsibility of local governments. Generally, however, much of the remaining costs are financed by local bonds that are paid back through a local tax allocation transferred from the national government. Therefore, ultimately, about 60 percent of initial project costs are financially supported by the national government while the remaining 40 percent are managed by local governments.

Operational costs for facilities are fully and directly covered by local governments. The two main revenue sources are the sale of designated plastic bags (a form of user fees in Japan) and general tax revenue.

Information and Human Resource Exchange

To promote connectivity and knowledge exchange between the national government and local governments, public officials and employees may take on roles in other levels of administration. There are also several mechanisms that allow local governments to report feedback to the national government, including the Japan Waste Management Association (JWMA), which includes 585 municipal governments, and the National Governors' Association. For example, at the annual meeting of the JWMA, local

Photo 7.1   **Japanese Bins**

Photo 7.2   **Japanese Recycling Facility**

governments submit feedback that is aggregated by the JWMA and shared with national agencies, including the Ministry of the Environment.

Japan's coordination in key dimensions of waste management ensures that best practices are disseminated across the country, planning is conducted in a data-driven manner, and cities have sufficient financial and human resources to process waste in a most sustainable manner.

5. Central Reforms to Stabilize the Waste Sector and Engage the Private Sector in Senegal

Senegal produces more than 2.4 million tonnes of waste per year. However, about 1.08 million tonnes remains uncollected. Of the waste that is collected, most is disposed of at a central dump that is one of the 10 largest dumpsites in the world. The country, which faces a rapid urbanization rate of 2.5 percent each year, has strongly focused on modernizing its waste management sector and developing the urban services needed by its burgeoning city population.

Although Senegal was interested in engaging the private sector to revitalize the waste management sector, it faced challenges typical of low- and middle-income countries related to transparency and difficulty in navigating the political system. Until 2015, waste management responsibilities were spread over several ministries, making coordination difficult. Furthermore, to invest in infrastructure and provide collection and disposal services, corporations require opportunities to recover costs. In Senegal, the lack of an established citizen payment system created financial gaps and led to payment delays that discouraged private entities.

Recognizing the pressing need to revitalize the waste sector, Senegal turned to internal reforms. The national government established a single public entity to streamline all waste management planning and services, called L'Unité de Coordination de la Gestion des Déchets Solides, or the Waste Coordination Unit, in 2015. This organizational structure was sustained even as regimes changed, and the government now has a mix of public and private service provision. The government structured a realistic relationship by devolving responsibilities to the private sector that are affordable to both the capital, Dakar, and the country at large. This structure is complemented by reliable and stable public entities that will follow through on contracts.

The waste management sector recovers 15 percent of operational costs, with the remaining 85 percent coming from the central government budget. Small, local private entities provide services from street cleaning to waste collection, and the government directly operates the remainder of the system. Waste is now collected daily in Dakar, streets are swept consistently, and most waste deposits have been cleaned up. The Waste Coordination Unit also began using media to communicate with citizens and optimized waste collection routes using web-based monitoring systems. They recruited young professionals to engage with modern technologies and implement progressive policies to ensure the long-term development of the waste sector.

The success of the new management structure has revived the interest of potential investors, including international donors.

The rapid improvement in waste service delivery in Senegal was made possible through radical changes in governance and improvements in technical capacity centrally. While Senegal has so far improved waste services without a traditional public-private partnership, the structural transformation in governance has created a more stable, attractive waste management sector for investors and waste management companies.

6. Decentralized Organic Waste Management by Households in Burkina Faso

In Burkina Faso, households historically managed waste through a traditional practice called *tampouré*. Tampouré involves storing organic waste in front of homes during the dry season and spreading the waste in fields before the first rains. The waste serves as a layer of nutritious compost and moisture-retaining mulch that improves agricultural production in areas of low productivity.

Currently, cities in Burkina Faso are growing rapidly along with waste and demand for agricultural products. To address these growing needs, the Ministry of Agriculture launched a Manure Pit Operation in 2001 that is in many ways inspired by the traditional practice of tampouré. Under this system, the government encourages households to establish pits and compost on their own land.

The government allocates funds each year to support household waste management. For example, between 2005 and 2012, the national government partnered with several development agencies to finance the construction of 15,000 manure pits in Burkina Faso's eastern region.

Currently, about 2 million tonnes of organic fertilizer is produced annually and used by farmers each year. A 2016 World Bank study revealed that 40 percent of the total waste produced by households in secondary cities and peri-urban areas in Burkina Faso was directly processed onsite (Banna 2017). This figure is remarkably high when compared with other parts of the African continent.

Burkina Faso's decentralized waste management system has significantly reduced the burden on the formal waste collection and disposal infrastructure. Its agricultural benefits have also led to increased food security and have created opportunities for citizens to generate income from waste.

7. Eco-Lef: A Successful Plastic Recycling System in Tunisia

Tunisia provides an example of successful integration of the informal recycling sector into waste management and of the application of the extended producer responsibility principle. In 1997, the Ministry of Environment launched a national program, Eco-Lef,[4] to address the significant issue of postconsumer packaging waste. The Eco-Lef program developed a national system for the recovery and recycling of postconsumer packaging primarily focused on plastic waste.

The Eco-Lef program is governed by a decree that specifies the methods required for the collection and management of bags and packaging waste (Republic of Tunisia 1997). The program is partly financed by the private sector through an eco-tax of 5 percent on the net added value of certain locally manufactured or imported plastic polymers.[5] The National Agency for Waste Management (ANGed) is responsible for administering the Eco-Lef program.

The Eco-Lef program has successfully improved postconsumer packaging collection and recycling rates. The system encourages individual and informal collectors to gather used plastic and metal packaging and deliver the materials to Eco-Lef collection centers. In return, waste collectors receive remuneration based on the type and quantity of packaging collected. There is a financial advantage for participating in the Eco-Lef system: prices for plastic packaging waste in a local market are about 500 dinars per tonne (US$208 per tonne) compared to 700 dinars per tonne (US$290/tonne) at Eco-Lef collection centers.

The system has an annual budget of US$5.8 million for 2018, and currently operates through 221 Eco-Lef collection centers, 41 of which are managed by ANGed and the remainder managed by the private sector (ANGed and Ministry of Social Affairs and Environment 2018). The centers have collected more than 150,000 tonnes of plastic packaging waste since the program's launch in 2001. Depending on the type of plastic, 70–90 percent of collected waste is recycled through more than 70 active private recyclers who receive plastic collected through the Eco-Lef system. Eco-Lef has contributed to the creation of about 18,000 jobs and 2,000 micro-enterprises for collection with the financial support of the National Employment Fund, a government fund that helps vulnerable populations find employment.

The Eco-Lef experience provides several key lessons:

- The extended producer responsibility principle can create a financially sustainable system for the collection, transportation, and recycling of materials.
- Government support in connection to legal, institutional, and operational activities is critical to the development of a recycling value chain.

Photo 7.3   Eco-Lef Workers Collecting and Weighing Packaging Waste at the Montplaisir Collection Center in Tunis, Tunisia

Source: Anis Ismail.

- Long-term ownership and management of the recycling system by the private sector can result in greater financial sustainability and operational efficiency.
- Integration of informal waste pickers into formal waste management operations can contribute to the success of recycling initiatives.

8. Extended Producer Responsibility Schemes in Europe

The European Union (EU) has integrated extended producer responsibility principles into its policies for more than 25 years (bio by Deloitte). The EPR landscape in the EU encompasses a large variety of schemes with different financial and technical configurations. Four EU directives set guidelines for specific waste streams, including packaging, end-of-life vehicles, batteries, and electronic equipment. Member states have the flexibility to develop specific regulations and operational mechanisms within EU guidelines (table 7.8.1).

Collective vs. Individual Compliance Schemes

Under the EU's EPR framework, producers may choose between a collective compliance scheme or an individual scheme. Under a collective scheme, individual legal obligations are outsourced to umbrella-type organizations, such as producer responsibility organizations (PROs). PROs are created to support producers in the handling of the technical, financial, and policy aspects of managing product life cycles. PROs receive financial contributions from industry and members and use these proceeds to recycle goods, manage data, conduct operations, facilitate contracting, and communicate with stakeholders. Under an individual scheme, producers that cater to a specific geography or that generate most of their waste close to the production site will manage waste directly, such as through a take-back program in which consumers can return used materials to the distributor.

The EU experience reveals that the most expensive schemes are not necessarily the best ones. Factors such as population density, citizen awareness, local laws, and legal frameworks also affect EPR performance. Furthermore, a country must consider recyclers in the informal market since formal EPR mechanisms reduce opportunities for them to collect materials.

Designing and implementing an EPR scheme involves a range of technical, financial, institutional, and legal considerations. A 2014 analysis of EU EPR

Table 7.8.1   Number of European Union Member States Implementing Extended Producer Responsibility Schemes in 2013

Legal framework	EPR scheme	Number of member states
Covered by specific European Union directives	Electrical and electronic equipment	28
	Batteries	28
	Packaging	26
	End-of-life vehicles	24
Not covered by European Union directives	Tires	20
	Graphic paper	11
	Oils	10
	Medical wastes, old and unused medicines	10
	Agricultural film	8

schemes identified four key pillars of success: (1) distribution of responsibilities across stakeholders, (2) recovery of true costs, (3) fair competition between PROs and operators, and (4) transparency by EPR schemes in reporting and transparency by the government in monitoring (bio by Deloitte 2014).

Distribution of responsibilities: Financial responsibility for a product's life cycle will often be borne by producers while the recycling programs themselves are operated by municipalities. At other times, producers will take a direct role in managing waste and contracting with private recyclers.

Recovery of true costs: EPR schemes must account for the costs of source segregation, collection, treatment, enforcement, and operation of the EPR program. In some cases, the amount that producers pay municipalities to recycle waste may depend on the final revenue generated in secondary materials markets, as is the case with the lubricant oil market in Germany. Governments may consider rewarding good product design with lower producer fees. In France, for example, graphic paper fees are calculated based on recyclability and other technical criteria. Finally, EPR profitability is tied to the performance of the secondary materials market, and legislators should plan EPR systems to be resilient across varying recycling markets.

Fair competition: A strong EPR system allows for competition between PROs and waste management operators. Competition encourages improvements in efficiency and reduces monopolies. Service operators should be procured using transparent procedures and competitive open tenders. PROs may be for-profit or nonprofit organizations, and are often owned by industry investors, such as within the battery industry in Austria and Denmark. An EPR system with free competition between PROs requires an independent body to verify compliance, centralize and aggregate performance reports, and ensure fair competition for all actors.

Transparency and monitoring: Monitoring the performance of an EPR system requires clear performance metrics such as unit costs and impact of the design on recycling activities. Metrics allow governments to compare the performance of different EPR schemes and support the replicability of good practices. EPR systems must also be monitored to reduce corruption, prevent lack of action, ensure that all waste is fully reported, optimize collection and treatment operations, and stay attuned to PRO activities and compliance. For example, Austria uses a two-tiered audit system to ensure the effective management of end products. Governmental authorities audit PROs, and PROs audit collection and treatment operators. Collective schemes can also be audited by their members.

EPR systems should adapt as new products are designed to ensure high recycling rates, minimal costs, and a strong transition to a circular economy.

Examples of producers' responsibilities within EPR schemes in the EU are as follows:

- Simple financial responsibility schemes in the United Kingdom, where producers are financing waste management operations
- Financial responsibility and partial organizational responsibility in Belgium, where producers provide financial compensation to

municipalities for collection while other activities, such as sorting, are fully managed within the private sector

- Full organizational responsibility where producers are contracting with private operators (electronic waste in France) or directly operating through collection and treatment of products (packaging waste in Germany)

In addition to operational, data management, and communication costs, fees might also cover the following:

- Contribution to a prevention fund (Austria, Belgium, and Czech Republic)
- Additional costs registered by municipalities such as use of public space or cleaning of container areas (Germany)
- Research and development programs or waste prevention activities (Austria, France, and Portugal)
- Litter prevention programs (Netherlands, Belgium)

Varying EPR models for treatment of end products include the following:

- *Packaging waste:* Direct management of end-products by producer (Czech Republic, France) or outsourcing to several PROs (7 in Austria, 10 in Germany, 39 in the United Kingdom)
- *End-of-life vehicles and oils:* Direct management of end products by producer (Germany) but more commonly outsourced to a single PRO (Finland, Italy, Portugal)
- *Electrical and electronic equipment and batteries:* Outsourced to a single PRO (the Netherlands, the Czech Republic) or several PROs (Austria, Denmark, the United Kingdom)

Photo 7.4 An Automated Bottle Deposit Machine

Source: Flaviu Pop.

9. Financially Resilient Deposit Refund System: The Case of the Bottle Recycling Program in Palau

Palau is a small country in the North Pacific with a population of 21,000 in 2015 (UN DESA 2014). Palau's economy relies on tourism visits to its famed Rock Islands and impressive diving sites. As of 2016, Palau received about 12,500 visitors per month.

Waste collection is coordinated within each state and waste disposal is the responsibility of the national Solid Waste Management Office of the Bureau of Public Works, which manages the M-dock semi-aerobic landfill, the country's largest landfill, situated in the capital city, Koror. Financially, solid waste management is funded entirely by the government. Although households and institutions are required to segregate waste streams, including for various recyclables and food waste, user fees are not charged or imposed on residents and businesses for waste collection and disposal, with the exception of a beverage container recycling program.

Solid waste generation is an increasing problem in Palau because of booming tourism and an increasing local population. Palau's waste system is inundated with food waste and plastics, composing 26 percent and 32 percent of waste, respectively. Tourism generates a large volume of beverage containers, and as an island state, plastic waste would overwhelm Palau if it is not addressed properly.

Palau's Beverage Container Recycling Program

In response to increasing plastic waste, the national government passed the Beverage Container Recycling Regulation in October 2006 to establish a national recycling program. The program is overseen by three main agencies: the Ministry of Finance (MOF), the Ministry of Public Infrastructures, Industries and Commerce (MPIIC), and the Koror State Government. The MOF manages the recycling fund, the MPIIC implements the recycling program, ensures sustainability, and identifies opportunities to export redeemed containers, and the Koror State Government operates a redemption center.

Palau's beverage recycling system addresses containers that are 32 ounces and smaller. The national government levies a US$0.10 deposit fee to consumers for plastic, glass, and metal containers, which are typically imported. When a container is returned to a redemption center, US$0.05 are returned to the customer, US$0.025 are channeled to Koror State, and the remaining US$0.025 are given to the national government to cover administrative costs.

The program began with a 6-month fundraising period to ensure operational sustainability during which beverage containers were taxed but the refund program was not yet in operation. This initial effort led to more than US$659,000 in revenue and funded the initial phases of the refund program. Through the program's full operation from 2011 to 2016, the national government earned US$2.2 million and refunded US$3.9 to customers by recovering more than 88 million imported containers. Through this system,

Photo 7.5   Compacting Beverage Containers inside the Plant in Palau

Source: Kevin Serrona.

about 8 percent of beverage containers were removed from the waste stream. In addition, about 98 percent of aluminum containers were recycled. Furthermore, about US$12,000 in operational costs were saved by diverting the containers from the M-dock semi-aerobic landfill. Collected beverage containers are shipped to Taiwan, China, for further processing.

The program has provided a variety of benefits. The national government has used profits to purchase heavy equipment to fix the slope of the M-dock semi-aerobic landfill, preventing a potential landslide of the waste. Koror State has used proceeds to buy balers and other equipment to improve the efficacy of the redemption center. The program has also provided employment to individuals who collect containers from other states.

Enabling Environment

Palau's beverage recycling program serves as a model for island countries that face limited space for waste management facilities and that possess sensitive natural environments. The program has operated sustainably because of the following factors:

- *Strong national government oversight:* The Recycling Act of 2006 effectively mandated a disposal fee for beverage containers. Because it is an island, the government maintains strong control over the entry of goods across its borders, which makes the program easier to monitor and manage.

- *Effective financial incentives:* The US$0.05 that residents are paid for each redeemed container makes the program sufficiently attractive to the public. The program has virtually removed used containers from the streets.
- *High public participation:* The key to the sustained operation of Palau's bottle recycling program is strong public support. Palauans realize the value of preserving their environment and the economic value of tourism.
- *Collaboration between Palau's national government and Koror State:* National and local collaboration in Palau was made possible by clear delineation of roles and responsibilities.

10. Improving Waste Collection by Partnering with the Informal Sector in Pune, India

The city of Pune has significantly advanced its solid waste management by entering into a public-private partnership with the organization SWaCH (Solid Waste Collection and Handling or, officially, SWaCH Seva Sahakari Sanstha Maryadit, Pune). SWaCH is India's first self-owned cooperative of waste pickers and other urban poor. In 2008, Pune Municipal Corporation (PMC) signed a five-year memorandum of understanding that gave SWaCH responsibility for collecting source-separated waste from households and commercial establishments, depositing the waste at designated collection points, and charging a user fee. The agreement also authorized waste collectors to retrieve and sell recyclables from aggregated waste.

Pune generates about 1,500–1,600 tonnes of solid waste per day. SwaCH provides door-to-door waste collection services to more than 500,000 households in the city and covers 60 percent of the geographical area. The remaining 40 percent not covered by SwaCH's collection operations either receive waste collection services directly from the city or dispose of waste in the city's community bins. The SwaCH door-to-door collection partnership has saved PMC about 510 million Indian rupees (about US$7.9 million) each year and has reduced carbon emissions significantly through reduced truck usage. In 2016, the agreement between PMC and SwaCH was renewed for another five years.

Overview of the Public-Private Partnership

Through the arrangement with PMC, 2,688 SwaCH members collect segregated waste from households, institutions, and businesses. Waste collectors sort dry waste in sheds provided by PMC and retrieve recyclables such as paper, glass, and plastic. Waste collectors retain all income from the sale of reclaimed materials, and in 2016 SwaCH diverted 50,000 tonnes of waste to recycling.

The door-to-door collection program was introduced first through a pilot in apartment complexes in wealthy areas, where citizens were highly aware, and had a willingness to pay, and were politically supportive of the initiative. The success of the pilot created demand in other areas of the city. Awareness initiatives, including rallies, one-on-one meetings, and political endorsement by local councilors, further generated support.

SWaCH members collect monthly user fees ranging from INR 10 to INR 40 (US$0.15–US$0.6) per household and INR 100 (US$1.5) per commercial entity for waste collection services. PMC partially subsidizes collection costs in slums so that households pay about INR 5 (US$0.07) per month. The total estimated cost of collection to the city is one of the lowest in India, at about INR 4.38 per month (US$0.06) in 2015.

SWaCH members also treat organic waste. Members are trained to operate biogas plants and to compost waste. To encourage citizens to treat waste at the source, PMC rebates 5 percent of property taxes to institutions that compost their own organic waste. Many of these institutions hire SWaCH members to collectively compost about 10 tonnes of waste per day. SWaCH members also operate biomethanation plants through build-operate-transfer contracts with the city.

Because the SwaCH model is based on customer satisfaction, the service provider is directly accountable to the user and has incentives to provide quality services. The service provider is entitled to collection of waste dumped outside of homes, which encourages user compliance. PMC conducts individual consultations with households to gain user support and levies penalties on users who fail to provide payment for services.

To provide financial resilience to the public-private partnership, PMC provides an ongoing annual grant to SWaCH that covers management and training costs, awareness-generation programs, and welfare benefits for members of SWaCH. The grant does not cover the salaries of collectors. Through the partnership with SWaCH, Pune has offered sustainable and efficient daily waste collection services to residents while improving the livelihoods of waste collectors within the city.

11. Improving Waste Management through Citizen Communication in Toronto, Canada

Toronto, Canada, uses citizen engagement to build a foundation for a more efficient solid waste management system. A multipronged communication strategy has been critical for reaching various residential audiences. Toronto has launched a detailed, interactive website that educates residents on garbage reduction, reuse, and recycling (City of Toronto 2018a). Waste management information that is relevant to citizens, such as source-separation guidelines, drop-off points, city regulations, and disposal rates and fees, are readily available on the site in a user-friendly, attractive manner.

Within the online platform, residents can use the Waste Wizard tool to understand how and on what day any item should be disposed of and when (figure 7.11.1). For instance, a search for items such as "pencil" and "clothes" yields advice on donating items in good condition wherever possible and disposing of the items in a garbage bin as a final option. A search for "plastic chair" results in guidance to place oversized items two feet away from the garbage bin on the next scheduled collection day (City of Toronto 2018b).

Toronto also actively uses social media to reach a wide audience. For instance, YouTube videos explain garbage to kids in a fun and simple way.[6] Videos are also available in foreign languages to reach growing populations living in multifamily homes, where recycling and composting rates (at 27 percent) are relatively lower than in single-family homes (at 65 percent) (McKay 2016).

Figure 7.11.1 Screenshot of Waste Wizard on the City of Toronto Website

Other successful initiatives to engage residents include a waste collection schedule mobile application and the 3Rs Ambassador Program, in which volunteers are trained to educate fellow residents on sustainable practices in waste reduction, reuse, and recycling. In 2016, the Mayor's Towering Challenge was organized to recognize notable reduce and reuse initiatives led by city residents. Toronto is currently focusing on applying these excellent communication strategies to implementation of its Long-Term Waste Strategy to achieve zero waste in the next 30 to 50 years.

12. Managing Disaster Waste

Depending on their nature and severity, disasters can create large volumes of debris ranging from 5 to 15 times the annual waste generation rates of the affected community (Reinhart and McCreanor 1999; Basnayake, Chiemchaisri, and Visvanathan 2006). The financial cost of managing disaster waste following a major event is also spiraling and has crossed the billion dollar mark in recent years (Thummarukudy 2012). For instance, after Hurricane Katrina, the cost of handling the disaster debris exceeded US$4 billion in a recovery effort that lasted more than three years (UNEP 2012).

The nature and composition of disaster waste differs based on the type of disaster and the built environment that has been affected. In most cases, the bulk of disaster waste is construction and demolition material such as concrete, steel, and wood. Disaster waste may also include natural debris such as trees, mud and rocks, food waste, damaged vehicles and boats, hazardous waste, and municipal waste.

With climate change modifying the frequency and intensity of many weather-related hazards (IPCC 2014), efficient, effective, and low-impact recovery is crucial (Brown 2015). If managed well, disaster waste can provide valuable resources for the postdisaster recovery and rebuilding process, generate income, and offset the use of virgin natural resources (Brown, Milke, and Seville 2011). Disaster waste can then be recycled, disposed of, used to generate energy, or repurposed for land reclamation and engineering fill (Brown 2015). Managing disaster waste well can help public agencies avoid future liabilities and costs, such as by recompacting unstable dumpsites or by cleaning up contaminated soil. Disaster waste is typically managed in three phases, detailed in table 7.12.1.

Japan is considered one of the most disaster-prepared countries in the world, and its plans and abilities were put to use in recent years (UNEP 2012).

Table 7.12.1   **Typical Phases of Disaster Waste Management**

Emergency response	Recovery	Rebuild
- Removal of debris and other immediate threats to public health and safety - Lasts between a few days and two weeks - Little scope for waste recycling and diversion	- Debris management as part of restoration efforts and building demolition - Majority of waste will be managed in this phase - Might be affected by factors beyond control of disaster waste managers such as returning residents - Can last for years (for example, Hurricane Katrina took five years)	- Debris management of wastes generated from, and used in, reconstruction - Longest phase; hard to define endpoint

Source: Brown, Milke, and Seville 2011.

Photo 7.6 **Recovery Efforts after Meethotamulla Dumpsite Collapse from Heavy Rains in Colombo, Sri Lanka**

In March 2011, a massive earthquake off the Pacific coast triggered a tsunami and damaged the Fukushima Daiichi Nuclear Power Plant, leading to the release of radioactive materials and the evacuation of thousands of people. Damages were estimated at more than US$210 billion, marking one of the most economically devastating disasters in history.

Japan's Ministry of the Environment formed a task force consisting of more than 100 experts from government agencies, research institutions, academia, and industry. Two months after the event, the Ministry of the Environment developed clear guidance notes for municipalities on dealing with disaster debris. The guidelines emphasized the importance of maximizing recycling opportunities and using local employment in recovery. Recognizing that many municipalities would be unable to handle the volume of disaster debris without assistance, the guidelines promoted collaboration between prefectures and jurisdictions. Additional funding was provided. These measures ensured consistency in the overall approach to the clean up, segregation, offsite transportation, and final disposal of debris.

Most middle- and lower-income countries that struggle to manage normal waste streams on a routine basis experience deeper crises after natural disasters. For instance, following the massive earthquake of April 25, 2015, in Nepal, waste accumulated in streets for several weeks. During that time, the need for clear guidelines for handling hazardous constituents such as paints and heavy metals became clear (UNEP 2015).

Over the past two decades, disaster waste management has been a growing focus of international policy-making and advisory efforts. Several guidance frameworks for public authorities have been developed, including *Disaster Waste Management Guidelines* developed by the Swedish Civil Contingencies Agency and the United Nations in 2011 (Joint UNEP/OCHA Environment Unit 2011).

13. Minimizing Food Loss and Waste in Mexico

In Mexico, more than 50 percent of waste is organic, a large share of which originates from food loss and waste (FLW). Food loss occurs during the production, storage, and distribution of food products, and food waste occurs when consumable food is thrown away. Of the many environmental consequences associated with FLW, a major impact is its contribution to greenhouse gas emissions (FAO 2013). In Mexico, conservative estimates from the Ministry of Environment and Natural Resources (2015) state that by 2020 the solid waste sector, including FLW, will be the fifth-largest source of greenhouse gas emissions in the country.

In 2016, the quantity, magnitude, and composition, as well as the environmental, social, and economic impacts of FLW in Mexico were comprehensively measured and assessed through a study supported by the World Bank (Aguilar Gutiérrez 2016). The study led to a legal reform that provided incentives for food donation and a call for the development of a national strategy. The following findings emerged from the study conducted in Mexico, and likely reflect similar situations in other countries in the region:

* More than 35 percent of total annual food production is lost or wasted at an annual economic cost of more than US$25 billion per year, or more than 2.5 percent of GDP.
* Greenhouse gas emissions from FLW are at least equal to that produced by 14 million cars per year.
* The estimated water loss associated with FLW is 39 billion liters per year.
* Estimated FLW in Mexico amounts to more than 20.4 million tonnes per year. The largest share of food loss occurs at the production stage, specifically during distribution and wholesale, while the largest share of food waste occurs at distribution centers and urban centers.
* More than 11 million people are living in extreme poverty in Mexico, often suffering from food insecurity, undernourishment, and malnourishment. The quantity of FLW exceeds the food requirements of this population.

Food Banks Are a Key Part of the FLW Solution

The Mexican Food Banking Network (BAMX) is a key organization responsible for promoting the reduction of FLW in the public and private sector's agenda and for motivating the recovery and channeling of food to combat hunger (BAMX 2015). BAMX began its operations as a nonprofit organization in 1995 to recover food to fight hunger and improve nutrition for vulnerable populations in Mexico.

One in four Mexicans suffers from food shortages and undernourishment, and since its inception, BAMX has effectively helped this

Photo 7.7   **Organic Waste Bin in Mexico City, Mexico**

vulnerable population while substantially reducing FLW. BAMX was one of the founding entities of the Global FoodBanking Network, an organization comprising 792 food banks in 32 countries. The Global FoodBanking Network is the second-largest network after that of the United States. BAMX is a key civil society organization in Mexico (World Bank 2016).

BAMX leads food rescue efforts resulting in savings of more than 120,000 tonnes annually and has benefited more than 1,137,000 people. It also purchases about 8,000 tonnes of food per year to achieve nutritional balance in the food donated to social causes.

In addition to supporting Mexico's efforts to fulfill the United Nations Sustainable Development Goals for 2030 (specifically, Goal 12.3 for

Responsible Consumption and Production), BAMX has also been a key advocate for several policies and legal proposals:

- In 2000, Mexico's General Law of Health was modified to encourage food donation.
- An amendment to the Income Tax Law provides tributary benefits to companies that donate their products five days before the expiration date.[7]
- As part of a presidential decree, companies that make food donations to food banks only receive an additional 5 percent fiscal incentive.

FLW and hunger are improving in Mexico as effective food banks are complemented by support from the government, the private sector, and international agencies.

14. Sustainable Source Separation in Panaji, India

Panaji is the capital of the state of Goa in southwest India, with a metropolitan area population of 114,759 according to the Census of India in 2011. Panaji is known for its strong cultural heritage and as a popular tourist destination, with colorful villas, hillside developments, and Portuguese influence. After the city's only landfill was closed in 2005, and faced with a vulnerable natural ecosystem and strong tourism economy, the city turned to sustainable practices. Source separation was the first step toward the vision of a landfill-free city, and today Panaji serves as a role model in solid waste management.

Source Separation Overview

Panaji generated 50 tonnes of waste daily in 2017. Residential waste is source separated into five streams through a system of colored bins:

- Green bins: Wet waste
- Black or grey bins: Glass and metals
- Pink bins: Paper and cartons
- Orange bins: Plastics
- White bins: Nonrecyclables

The City Corporation of Panaji provides door-to-door collection of waste. Wet waste is collected from households every day, and dry waste is collected twice a week. Household wet waste is composted at one of 96 decentralized compost units, whereas wet waste from commercial establishments is treated using windrow composting at two bulk processing units, and it is used for urban horticulture projects. In some cases, wet waste from hotels is digested onsite to produce biogas. In total, the city processes about 24 tonnes of wet waste daily.

After collection, the city's dry waste is stored and aggregated at one of 12 sorting units. From there, the dry waste is further segregated into 20 different streams at one of two Material Recycling Facilities. Each day, about 4 tonnes of dry waste is sent to a cement processing plant in Wadi, Karnataka, and 3 tonnes of recyclables are auctioned to vendors. In 2016, about US$22,000 in revenue was generated from the sale of recyclables.

Hazardous waste, such as batteries and tube lights, is also separated and processed at a specialized treatment facility.

Success Factors

The success of Panaji's source separation program can be attributed to strong public engagement, financial management, and institutional commitment.

- *Public engagement:* The city promoted sustainable waste practices through a public campaign called "Bin Free in 2003." The city engaged local students, celebrities, business leaders, and neighborhood civic

Photo 7.8a and b   Sorting Center at Residential Colony in Panaji, India

Photo 7.8c   Decentralized Composting Units in Panaji, India

Source: Ritu Thakur.

bodies to promote source segregation. As part of the campaign, the city spearheaded cultural programs ranging from music festivals to carnivals to encourage citizens to take responsibility for city cleanliness. Finally, the city launched a program called "Waste Wise" in all schools to promote waste segregation and to provide incentives for environmentally friendly behavior.

- *Pilot design:* The city's source separation program was first launched at a pilot scale covering only 70 households and two waste streams, one for dry waste and one for wet waste.
- *Enforcement mechanisms:* To encourage citizen engagement, the city removed all community waste bins, which required households to manage their waste privately. Simultaneously, the city introduced a door-to-door collection program through which households must personally hand their waste to collectors. Paid sanitation workers inspect waste, and a combination of the personal exchange with households and formal daily monitoring motivates compliance.
- *Financial management:* The City Corporation of Panaji has achieved financial sustainability for the solid waste program through fees and recycling revenues. The city established a new sanitation fee for households as well as a higher commercial fee for institutions. Households are charged a flat fee of INR 500 (US$7.3) for door-to-door collection in combination with the property tax, and commercial entities are charged between INR 600 (US$8.7) and INR 11,000 (US$16). User fees are strongly enforced with penalties. Some remaining costs are subsidized by city funds.
- *Institutional commitment and responsiveness:* The city government formed a new Solid Waste Management department that is overseen by a Waste Management Officer. The program also created a centralized complaint redress system, which includes a 24-hour helpline for unattended garbage and a dedicated vehicle to quickly respond to urgent waste situations. Finally, the city ensured that workers were provided with healthy and safe working conditions.

15. Musical Garbage Trucks in Taiwan, China

Three decades ago, the waste disposal system in Taiwan, China, looked nothing like it does today. Garbage collection spots were overflowing, smelly, and infested with rats and insects. Most waste was disposed of in dumpsites. In the late 1980s, the Taiwanese government decided to revitalize its waste system by implementing strict guidelines and regulations and promoting recycling. A circular economy emerged as a new guiding principle.

Today, Taiwan, China is a leader in recycling, with its Environmental Protection Administration reporting a 55 percent recycling rate in 2015. A key component of its comprehensive strategy is motivating community involvement through popular "musical" garbage trucks (Shen n.d.).

On this small, densely populated island, most families live in apartments and do not have communal garbage bins close to their homes. Within their households, residents are required to separate their waste into three categories—general (or nonrecyclable) waste recyclables, and food waste. Purchase of a blue City of Taipei garbage bag for general refuse is also compulsory in the capital. These bags are available at most corner stores and come in differently priced sizes, ranging from 3 to 120 liters. Recycling, on the other hand, is free. This encourages citizens to recycle and produce less trash.

At night, garbage trucks alert people to their arrival with high-pitched, familiar tunes, such as Beethoven's "Für Elise." Residents gather on the streets with their bags of general waste and throw them into the yellow musical trucks. Three times a week, recyclables are collected in a separate white pickup truck that follows the yellow garbage truck. Volunteers and officials stand on the back of the truck to help citizens sort their recyclables correctly (Bush 2017). This way, trash is delivered straight from home to truck without ever touching the ground.

The trucks run on a regular schedule so residents are ready with their bags of waste when the curbside melodies begin. In the capital, Taipei, there are more than 4,000 pickup spots over five nights each week. Residents can receive alerts on nearby stops via mobile phone applications.

Once collected, most general waste is incinerated, raw food waste is converted to fertilizer for farmers, cooked food waste is used as livestock feed, and recyclables are sorted and recycled. Public authorities monitor compliance using video cameras and financially reward citizens that report misbehavior. Violators are fined and may be publicly chastised for their offense.

16. The Global Tragedy of Marine Litter

Across the world, beaches and waterways scattered with litter are an increasingly common sight and this marine litter has serious impacts on the environment, public health, and the economy.

Marine litter comes in all shapes and sizes and, depending on the material, could be damaging to human health. Some 90 percent of floating marine debris is plastic, of which nearly 62 percent is food and beverage packaging (Galgani, Hanke, and Maes 2015; Consultic 2013). Although plastics have been mass-produced for only about 60 years, they persist in open waters for decades and even centuries (Andrady 1994). Even plastics designed to be biodegradable may not fully decompose since they depend on factors such as exposure to light, oxygen, and temperature (Swift and Wiles 2004), which are scarce in ocean depths. Smaller particles of plastic from manufacturing processes could also be difficult to account for and nearly impossible to extract.

Marine litter can be land- or sea-based and often results from poor solid waste management practices. An estimated 80 percent of marine litter originates from land-based sources such as mismanaged dumps and landfills, storm water discharge, sewage, industrial facilities, and coastal tourism (Arcadis 2014; McIlgorm, Campbell, and Rule 2008). Waste may also be transported to the ocean from inland rivers. In 2010, an estimated 32 million tonnes of plastic waste were mismanaged in coastal areas, allowing between 4.8 and 12.7 million tonnes of plastic waste to escape into oceans (UNEP and NOAA 2012; Jambeck et al. 2015). When collection systems and disposal sites are in proper operation, waste is less likely to be disposed of haphazardly.

Photo 7.9   **Spilled Garbage on the Beach**

Although research is in its infancy, sufficient evidence indicates that marine litter has a detrimental effect on society. One study estimates that costs associated with ocean-based plastic consumer waste leads to losses of US$8 billion annually, including revenue losses to fisheries, aquaculture, and marine tourism industries in addition to the cost of cleaning up litter on beaches (UNEP 2014).

Marine debris affects marine life through debris entanglement, which injures marine life or makes escaping for air or consuming food impossible (Laist 1997). Marine litter can also be ingested by sea organisms, with negative effects on reproduction and development for both the organisms themselves and downstream consumers. A study revealed that marine litter was present in all marine turtles studied, 59 percent of whales, 36 percent of seals, and 40 percent of seabirds (Foekema et al. 2013). Plastic particles have even been found in many species of fish and shellfish sold for human consumption.

Marine waste is expected to grow with increasing population and rising per capita consumption, especially in urban areas and quickly developing economies. Several policy initiatives related to marine environmental protection and pollution have been drafted along with action plans at the regional, national, and municipal levels to address the problem. The *UN 2030 Agenda for Sustainable Development,* adopted in 2015, provides an overarching framework to guide international, regional, national, and local initiatives. Four out of the 17 Sustainable Development Goals have associated targets particularly relevant to marine plastic pollution. At the national level, Japan, the Republic of Korea, the Netherlands, and Singapore have developed legislation and policies to address marine litter, but such legislation remains uncommon globally. At a municipal level, many cities are improving waste management practices, and some are implementing plastic bans or penalties on bottles and bags, which can reduce plastic usage and waste if enforced.

17. Using Information Management to Reduce Waste in Korea

Korea is a high-income country with a population of almost 50 million people in 2015 (UN DESA 2014). The country generated 18.2 million tonnes of municipal waste in 2014 and recycled 58 percent of its waste, achieving one of the highest rates among Organisation for Economic Co-operation and Development countries (Kho and Lee 2016). Korea is also a global leader in progressive solid waste management legislation, under which residents are encouraged to recycle plastics through a deposit refund system, reduce waste through volume-based fees, and separate different streams of waste at the household and business levels. The country has also established an extended producer responsibility system and recycles its construction waste.

Korea established its solid waste management information system in 2001. The system digitally records statistics from waste generation to transportation and final disposal. The system was developed by the Ministry of Information and Communication and comprises three separate databases. The Allbaro System is an overarching platform that monitors all waste transportation and disposal activities. The system tracks the amount of waste that is collected and transported, analyzes truck routes using a geographic information system, records licenses and authorization documents, and aggregates statistics to support evidence-based policy making. The second system uses radio-frequency identification (RFID) to obtain information about food waste management. Each household uses a personalized card containing an RFID chip to open local food waste disposal bins. Through this process, the identity of the household and the weight of waste disposed of is instantly recorded, and households are charged a fee based on waste amount. In 2013, a 20 percent reduction in food waste was achieved in the capital city of Seoul through this digital volume-based system. Finally, the Recyclable Information System matches suppliers and buyers of recycled materials on a centralized platform. The platform provides information on recycling prices and technologies, matches businesses to recycling firms, and facilitates electronic bidding. As of September 2013, the platform had registered 69,000 members.

Korea's waste information system has led to cost savings, promoted transparency, and eliminated illegal waste disposal. The phased implementation of the system allowed stakeholders to adjust to the system successfully. Other success factors include a public relations strategy that increased awareness and a strong feedback mechanism that enabled the system to be improved over time. Most importantly, the waste information system allowed a volume-based waste fee to be implemented, which has led to a valuable change in citizen behavior and reduced resource consumption.

Notes

1. In 2015, it became the Ministry of Environment and Sustainable Development.
2. https://cursos.ambiente.gob.ar/sigirsu/.
3. Entities established by several municipalities and wards to jointly conduct administrative services.
4. "Eco" for ecology and "Lef" for packing in Arabic.
5. Such as ethylene products, propylene, styrene, vinyl chloride, and vinyl acetate.
6. Videos are available on the City of Toronto's YouTube channel: https://www.youtube.com/watch?v=aNrX4MZxszo&list=PLp11YxteHNp25DJbbfEeuY59Z3gPPOtdg.
7. Companies that produce or commercialize food are exempt from any liability for possible damage to the health of third parties caused by food donated through food banks, and food banks are responsible for the good management of donated products.

References

Aguilar Gutiérrez, G. 2016. "Food Losses and Food Waste in Mexico: Quantification and Some Proposals for Public Policy." International Workshop on Food Loss and Food Waste, Washington, DC, November 7–9.

Andrady, A. L. 1994. "Assessment of Environmental Biodegradation of Synthetic Polymers." *Journal of Macromolecular Science Part C Polymer Reviews* 34 (1): 25–76.

ANGed and Ministry of Social Affairs and Environment. 2018. Eco-Lef Progress Report, February 22, 2018, in Arabic Power Point format. https://www.linkedin.com/in/anged-tunisie/detail/recent-activity/shares.

Arcadis. 2014. "Marine Litter Study to Support the Establishment of an Initial Quantitative Headline Reduction Target." SFRA0025. European Commission DG Environment Project number BE0113.000668.

BAMX (Bancos de Alimentos de Mexico). 2015. *Informe 2015.* (online). https://bamx.org.mx/wp-content/uploads/2016/09/Informe-Anual-2015.pdf.

Banna, Farouk. 2017. "Municipal Solid Waste Management in Burkina Faso: Diagnostic and Recommendations." World Bank, Washington, DC.

Basnayake B. F. A., C. Chiemchaisri, C. Visvanathan. 2006. "Wastelands: Clearing Up after the Tsunami in Sri Lanka and Thailand." *Waste Management World* 31–38.

bio by Deloitte. 2014. *Development of Guidance on Extended Producer Responsibility (EPR). Final Report.* Report prepared for the European Commission. Neuilly-sur-Seine: bio by Deloitte.

Brown, Charlotte. 2015. "Waste Management Following Earthquake Disaster." In *Encyclopedia of Earthquake Engineering*, edited by M. Beer, E. Patelli, I. Kougioumtzoglou, 3921–34. Berlin, Heidelberg: Springer.

———, Mark Milke, and Erica Seville. 2011. "Disaster Waste Management: A Review Article." *Waste Management* 31: 1085–98.

Bush, Jessica. 2017. "Taiwan Has Found a Brilliant Way to Get People to Recycle More." Buzzfeed, August 30. https://www.buzzworthy.com /taiwan-garbage-disposal/.

City of Toronto. 2018a. "Recycling, Organics & Garbage." https://www .toronto.ca/services-payments/recycling-organics-garbage/.

City of Toronto. 2018b. "Waste Wizard." https://www.toronto.ca/services -payments/recycling-organics-garbage/waste-wizard/.

Cohen, Anna. 2012. "‏הנביב הביבסה ןעמל קורי דעצ - יטאמואנפ הפשא יוניפ.‏ ‏הביבסהו‏" ("Pneumatic garbage disposal—a green step for the environment in Yavneh and the surrounding area.") MYavne. July 25.

Consultic. 2013. "Post-Consumer Plastic Waste Management in European Countries 2012 - EU 27 + 2 Countries." Consultic, Rückersdorf, Germany. http://kunststofkringloop.nl/wp-content/uploads/2014/05/Plastic-waste -management-report-October-2013-versie-NL-en-EU-voor-keten akkoord.pdf.

Correal, Magda. 2015. "Estrategia Nacional para el desarrollo de infraestructura." Sector Aseo. Bogota, Colombia.

———. 2016. "Modelos de regionalización y mecanismos de recuperación de costos de la prestación del servicio." Foro Internacional "Organismos Operadores para la Gestión Integral de los Residuos Sólidos Urbanos." IBD-SEMARNAT. Mexico City, April 20–21.

Economist Intelligence Unit. 2011. *US and Canada Green City Index—Assessing the Environmental Performance of 27 Major US and Canadian Cities.* Munich: Siemens, A.G. https://www.siemens.com/entry/cc/features /greencityindex_international/all/en/pdf/report_northamerica_en.pdf.

EPA (United States Environmental Protection Agency). 2017. "Zero Waste Case Study: San Francisco." Managing and Transforming Waste Streams—A Tool for Communities. https://www.epa.gov/transforming -waste-tool/zero-waste-case-study-san-francisco.

FAO (Food and Agriculture Organization of the United Nations). 2013. *Food Wastage Footprint—Impacts on Natural Resources* (online). http:// www.fao.org/docrep/018/i3347e/i3347e.pdf.

Foekema, E. M., C. De Gruijter, M. T. Mergia, J. A. van Franeker, A. J. Murk, and A. A. Koelmans. 2013. "Plastic in North Sea Fish." *Environmental Science and Technology* 47 (15): 8818–24.

Galgani, Francois, Georg Hanke, and Thomas Maes. 2015. "Global Distribution, Composition and Abundance of Marine Litter." In *Marine Anthropogenic Litter,* edited by Melanie Bergmann, Lars Gutow, and Michael Klages, 29–56. London: Springer.

IPCC (Intergovernmental Panel on Climate Change). 2014. *Climate Change 2014: Impacts, Adaptation, and Vulnerability.* New York, and Cambridge, UK: Cambridge University Press. http://www.ipcc.ch/report /ar5/wg2/.

Jambeck, Jenna R., Roland Geyer, Chris Wilcox, Theodore R. Siegler, Miriam Perryman, Anthony Andrady, Ramani Narayan, and Kara Lavender Law. 2015. "Plastic Waste Inputs from Land into the Ocean." *Science* 347 (6223): 768–71. https://www.iswa.org/fileadmin/user_upload/Calendar _2011_03_AMERICANA/Science-2015-Jambeck-768-71__2_.pdf.

Kho, Pan-Ki, and Syung-Uk Lee. 2016. "Waste Resources Management and Utilization Policies of Korea." Korea Research Institute for Human Settlements, Publications Registration Number 11-10510000-000755 -01, Republic of Korea. Accessed May 27, 2017. http://www.ksp.go.kr /publication/modul.jsp.

Laist, David W. 1997. "Impacts of Marine Debris: Entanglement of Marine Life in Marine Debris Including a Comprehensive List of Species with Entanglement and Ingestion Records." In *Marine Debris*, edited by J. M. Coe and D. B. Rogers, 99–139. New York: Springer Series on Environmental Management.

McIlgorm, A., H. F. Campbell, and M. J. Rule. 2008. *Understanding the Economic Benefits and Costs of Controlling Marine Debris in the APEC Region.* (MRC 02/2007). A report to the Asia-Pacific Economic Cooperation Marine Resource Conservation Working Group by the National Marine Science Centre (University of New England and Southern Cross University), Coffs Harbour, NSW, Australia, December.

McKay, Jim. 2016. "City of Toronto Solid Waste Management Services." Presented at "Property Management Expo," November 30. https:// ccitoronto.org/sites/default/uploads/files/Jims-waste-presentation -Jan2017.pdf.

Ministry of Environment and Natural Resources (Secretaría de Medio Ambiente y Recursos Naturales). 2015. Informe de la Situación del Medio Ambiente en México.

Nakou, D., A. Benardos, and D. Kaliampakos. 2014. "Assessing the Financial and Environmental Performance of Underground Automated Vacuum Waste Collection Systems." *Tunneling and Underground Space Technology* 41: 263–71. https://www.researchgate.net/publication/260114762_Assessing _the_financial_and_environmental_performance_of_underground _automated_vacuum_waste_collection_systems.

Reinhart, Debra R., and Philip T. McCreanor. 1999. *Disaster Debris Management—Planning Tools.* Washington, DC: US Environmental Protection Agency. https://www.researchgate.net/publication/237778695 _Disaster_Debris_Management_-_Planning_Tools.

Republic of Tunisia. Presidency of the Government. National Portal of Legal Information. Decree no 97-1102 of 2 Juin 1997. http://legislation.tn.

Revolvy. n.d. "Automated Vacuum Collection." https://www.revolvy.com/main/index.php?s=Automated%20vacuum%20collection.

Shen, Stephen. n.d. "Waste Management Policies and Services in Taipei." https://www.pecc.org/resources/infrastructure-1/1246-towards-zero-waste-society-new-management-policies-for-solid-waste-disposal-in-chinese-taipei-1/fil. p. 81–88.

Swift, G., and D. Wiles. 2004. "Biodegradable and Degradable Polymers and Plastics in Landfill Sites." *Encyclopedia of Polymer Science and Technology*, edited by J. I. Kroschwitz. Hoboken: John Wiley & Sons.

Thummarukudy, Muralee. 2012. "Waste: Disaster Waste Management: An Overview." In *Environment Disaster Linkages (Community, Environment and Disaster Risk Management*, Volume 9, edited by Rajib Shaw and Phong Tran, 195–218. Bingley, UK: Emerald Group Publishing Limited.

UN DESA (United Nations Department of Economic and Social Affairs). 2014. "World Urbanization Prospects: The 2014 Revision." CD-ROM Edition. UN DESA, Population Division, New York.

UNEP (United Nations Environmental Programme). 2012. *Managing Post-Disaster Debris: The Japan Experience*. Geneva: United Nations Environmental Programme.

———. 2014. *Valuing Plastics: Business Case for Measuring, Managing and Disclosing Plastic Use in the Consumer Goods Industry*. Nairobi: United Nations Environment Programme.

———. 2015. *Nepal Final Draft. Disaster Waste Management Policy, Strategy & Action Plan*. Kathmandu, Nepal: UNEP.

——— and NOAA. 2012. *The Honolulu Strategy—A Global Framework for Prevention and Management of Marine Debris*. Nairobi: United Nations Environment Programme; and Washington, DC: National Oceanic and Atmospheric Administration. https://marinedebris.noaa.gov/sites/default/files/publications-files/Honolulu_Strategy.pdf.

UNEP-UNOCHA. 2011. *Disaster Waste Management Guidelines*. Geneva: Emergency Preparedness Section, Joint UNEP/OCHA Environment Unit.

Additional Resources

Achieving Financial Sustainability in Argentina and Colombia

Solda, S., I. Berardo, and M. Mosteirin. 2014. "Integrated Urban Solid Waste Management Economic and Financial Matrix. A Methodology for Calculating Integrated Solid Waste Management Cost: The Case of Argentina." National Solid Waste Management Project Implementation Unit. National Secretariat of Environment and Sustainable Development. Paper presented at the International Solid Waste Association Conference, São Paulo, Brazil.

Economic and Financial Sustainability in Municipalities. Argentina. Presentation by the Executing Unit of Argentina—National Secretariat of Environment and Sustainable Development at the International SWM Workshop in Washington, DC, May 2013.

Information provided by Santiago Solda through several e-mail messages.

https://cursos.ambiente.gob.ar/sigirsu/.

http://www.cra.gov.co/es/acerca-de-la-entidad/estructura-organizacional.

Comisión de Regulación de Agua Potable y Saneamiento Básico. 2015. Nuevo marco tarifario del servicio público de aseo para municipios con más de 5.000 suscriptores en el área urbana.

Automated Waste Collection in Israel

Alon Group. 2018. "The State of Garbage and Waste in Israel." http://www .alon-group.com/en/the-state-of-garbage-and-waste-in-israel/.

Central Reforms to Stabilize Waste Sector and Engage Private Sector in Senegal

AllAfrica. 2016. "Senegal: Solid Waste Management in Dakar—Ucg gets a new tool." https://translate.googleusercontent.com/translate_c?depth=1& hl=en&prev=search&rurl=translate.google.com&sl=fr&sp =nmt4&u=http://fr.allafrica.com/stories/201606171199.html&usg=ALkJ rhiIQ3vvgmsZHsB3XSQ03WvwgFA8rA.

World Bank. 2017. "Project Information Document/Integrated Safeguards Data Sheet (PID/ISDS)." World Bank, Washington, DC. http://documents .worldbank.org/curated/en/581531500995135875/pdf/ITM00184 -P161477-07-25-2017-1500995132357.pdf.

Decentralized Organic Waste Management by Households in Burkina Faso

International Monetary Fund. 2005. "Burkina Faso: Poverty Reduction Strategy Paper." Country Report 05-338. Washington, DC. https://www .imf.org/external/pubs/ft/scr/2005/cr05338.pdf.

World Bank. 2005. "Burkina Faso Joint IDA-IMF Staff Advisory Note on the Poverty Reduction Strategy Paper and the Annual Progress Report of the Poverty Reduction Strategy Paper." Report No. 31749-BF. Washington, DC. http://documents.worldbank.org/curated/en/42594146 8239680045/text/31749.txt.

Financially Resilient Deposit Refund System: The Case of the Bottle Recycling Program in Palau

Bureau of Immigration, Ministry of Justice, Bureau of Planning and Ministry of Finance, Republic of Palau: 2016. http://palaugov.pw/immigration -tourism-statistics/.

ADB (Asian Development Bank). 2014. "Solid Waste Management in the Pacific: Palau Country Snapshot." ADB Publication Stock No. ARM1466 11-2, Manila.

JICA (Japan International Cooperation Agency) and Palau Ministry of Resources and Development). 2008. "Draft National Solid Waste Management Plan." JICA, Palau.

Improving Waste Collection by Partnering with the Informal Sector in Pune, India

https://swachcoop.com/.

Bhaskar, A., and P. Chikarmane. 2012. "The Story of Waste and Its Reclaimers: Organising Waste Collectors for Better Lives and Livelihoods." *The Indian Journal of Labour Economics* 55 (4): 595–619. https://swachcoop.com/pdf/AnjorBhaskar.pdf.

Chikarmane, P. 2012. "Integrating Waste Pickers into Municipal Solid Waste Management in Pune." WIEGO. https://swachcoop.com/pdf /SWaCH%20policy%20brief.pdf.

Managing Disaster Waste

Ekici, Siddik, David A. McEntire, and Richard Afedzie. 2009. "Transforming Debris Management: Considering New Essentials." *Disaster Prevention and Management: An International Journal* 18 (5): 511–22.

UNISDR (United Nations International Strategy for Disaster Reduction). 2007. *Hyogo Framework for Action 2005–2015: Building the Resilience of Nations and Communities to Disasters.* Geneva: United Nations Office for Disaster Risk Reduction. https://www.unisdr.org/we/coordinate/hfa.

———. 2015. *The Sendai Framework for Disaster Risk Reduction 2015– 2030.* Geneva: United Nations Office for Disaster Risk Reduction.

UNEP (United Nations Environmental Programme). 2005. *Lessons Learnt from the Tokage Typhoon (Typhoon 23 of 2004) in Japan.* Geneva: United Nations Environmental Programme.

US EPA (United States Environmental Protection Agency). 2008. *Planning For Natural Disaster Debris.* Washington, DC: Office of Solid Waste and Emergency Response and Office of Solid Waste.

Minimizing Food Loss and Waste in Mexico

Secretaría de Medio Ambiente y Recursos Naturales. *El medio Ambiente en Mexico 2013–2014* (online). http://apps1.semarnat.gob.mx/dgeia/informe _resumen14/05_atmosfera/5_2_2.html.

International Workshop on Food Loss and Food Waste. Washington, DC. 2016. Federico Gonzalez Celaya (online). http://www.worldbank.org/en /events/2016/11/07/2016-international-workshop-on-food-loss-and -food-waste (July 10, 2016).

Sustainable Source Separation in Panaji, India

Corporation of Panaji, http://ccpgoa.com/index.php.

The Water and Sanitation Programme. 2006. "Panaji (Goa): Innovation and Incentives Work Wonders." In *Solid Waste Management Initiatives in Small Towns—Lessons and Implications,* 11–17. Washington, DC: World Bank. http://ccpgoa.com/swachhbharat/uploads/download/18 _down_SWM_initatives_in_small_towns-Case_Study.pdf.

The Global Tragedy of Marine Litter

Cole, M., P. Lindeque, E. Fileman, C. Halsband, and T. S. Galloway. 2015. "The Impact of Polystyrene Microplastics on Feeding, Function and Fecundity in the Marine Copepod Calanus Helgolandicus." *Environmental Science and Technology* 49 (2): 1130–37.

Davison, P., and R. G. Asch. 2011. "Plastic Ingestion by Mesopelagic Fishes in the North Pacific Subtropical Gyre." *Marine Ecology Progress Series* 432 (June): 173–80.

European Commission. 2010. "Marine Litter: Time to Clean Up Our Act." European Commission, Brussels. http://ec.europa.eu/environment /marine/pdf/flyer_marine_litter.pdf.

———. 2012. "Overview of EU Policies, Legislation and Initiatives Related to Marine Litter." SWD(2012) 365 final. European Commission, Brussels. http://ec.europa.eu/environment/marine/pdf/SWD_2012_365.pdf.

Gall, S., and R. Thompson. 2015. "The Impact of Debris on Marine Life." *Marine Pollution Bulletin* 92 (1-2): 170–79.

GESAMP (Joint Group of Experts on the Scientific Aspect of Marine Environmental Protection). 2015. *Sources, Fate and Effects of Microplastics in the Marine Environment: A Global Assessment.* London: International Maritime Organization. http://ec.europa.eu/environment /marine/good-environmental-status/descriptor-10/pdf/GESAMP_micro plastics%20full%20study.pdf.

Meeker, J. D., S. Sathyanarayana, and S. H. Swan. 2009. "Phthalates and Other Additives in Plastics: Human Exposure and Associated Health Outcomes." *Philosophical Transactions of the Royal Society of London B: Biological Sciences* 364 (1526): 2097–113.

Moss, E., A. Eidson, and J. Jambeck. 2017. *Sea of Opportunity: Supply Chain Investment Opportunities to Address Marine Plastic Pollution.* New York: Encourage Capital on behalf of Vulcan.

Napper, I. E., A. Bakir, S. J. Rowland, and R. C. Thompson. 2015. "Characterisation, Quantity and Sorptive Properties of Microplastics Extracted from Cosmetics." *Marine Pollution Bulletin* 99 (1–2):178–85.

Rochman, Chelsea M., Akbar Tahir, Susan L. Williams, Dolores V. Baxa, Rosalyn Lam, Jeffrey T. Miller, Foo-Ching Teh, Shinta Werorilangi, and Swee J. Teh. 2015. "Anthropogenic Debris in

Seafood: Plastic Debris and Fibers from Textiles in Fish and Bivalves Sold for Human Consumption." *Scientific Reports 5*, Article number 14340, https://www.nature.com/articles/srep14340.

Sheavly, S. B. 2007. "National Marine Debris Monitoring Program: Final Program Report, Data Analysis and Summary." Prepared for U.S. Environmental Protection Agency by Ocean Conservancy. http://www .scirp.org/(S(i43dyn45teexjx455qlt3d2q))/reference/ReferencesPapers .aspx?ReferenceID=1692635.

UNEP (United Nations Environment Programme). 2005. *Marine Litter: An Analytical Overview*. Nairobi: United Nations Environment Programme. http://www.cep.unep.org/content/about-cep/amep/marine-litter-an-ana lytical-overview/view.

———. 2016. "Marine Plastic Debris and Microplastics: Global Lessons and Research to Inspire Action and Guide Policy Change." Nairobi: United Nations Environment Programme. http://drustage.unep.org /about/partnerships/marine-plastic-debris-and-microplasticsglobal -lessons-and-research-inspire-action-and-guide-policy.

——— and GRID-Arendal. 2016. *Marine Litter Vital Graphics*. Nairobi and Arendal: United Nations Environment Programme and GRID-Arendal. https://www.grida.no/publications/60.

Wright, S. L., R. C. Thompson, and T. S. Galloway. 2013. "The Physical Impacts of Microplastics on Marine Organisms: A Review." *Environmental Pollution* 178: 483–92.

Waste Generation (tonnes per year) and Projections by Country or Economy

Country or economy	Region	Income	Original year reported			Source
			MSW generation	Population	Year	
Afghanistan	SAR	LIC	5,628,525	34,656,032	2016	World Bank 2016b, 18
Albania	ECA	UMIC	1,142,964	2,880,703	2015	Albania, INSTAT 2016, 2
Algeria	MENA	UMIC	12,378,740	40,606,052	2016	Ouamane 2017
American Samoa	EAP	UMIC	18,989	55,599	2016	SPREP 2016, 21
Andorra	ECA	HIC	43,000	82,431	2012	UNSD 2016
Angola	SSA	LMIC	4,213,644	25,096,150	2012	Angola, Ministry of Environment 2012, 6
Antigua and Barbuda	LAC	HIC	136,720	98,875	2009	Caribbean Community Secretariat 2013, 147; Francis et al. 2015
Argentina	LAC	UMIC	17,910,550	42,981,515	2014	World Bank 2015b, 35
Armenia	ECA	LMIC	492,800	2,906,220	2014	Armenia, National Statistical Service 2017
Aruba	LAC	HIC	88,132	103,187	2013	Pricewaterhouse Coopers Aruba 2014, 11
Australia	EAP	HIC	13,345,000	23,789,338	2015	OECD 2018
Austria	ECA	HIC	4,836,000	8,633,169	2015	Eurostat 2017
Azerbaijan	ECA	UMIC	2,930,349	9,649,341	2015	Azerbaijan, Ministry of Economy 2017, 80 (table 4.5)

Comment	2016 adjusted		2030 projected		2050 projected	
	MSW generation	Population ('000s)	MSW generation	Population ('000s)	MSW generation	Population ('000s)
1 (Kabul, 0.7 kg/person/day)	5,628,525	34,656	7,979,843	46,700	12,887,446	61,928
	1,178,111	2,926	1,320,644	2,933	1,392,409	2,664
2	12,378,740	40,606	16,319,973	48,822	21,171,891	57,437
1 (urban, 1 kg/person/day; rural regional average, 0.5 kg/person/day)	18,989	56	21,468	57	25,433	57
3	43,594	77	45,675	78	49,509	77
4 (0.46 kg/person/day)	4,829,098	28,813	7,668,976	44,712	13,468,138	76,046
Municipal waste collected from HH and deposited in the landfill is 22,700 tonnes in 2009. HH generation is adjusted for the amount of uncollected household waste by dividing HH waste generation by the fraction of households with waste collection in 2011 (0.9861). Municipal waste collected from other origins is added (113,700 tonnes) from 2009.	33,239	101	64,920	115	79,530	125
4 (49.07 tonnes/day)	18,184,606	43,847	23,740,083	49,323	31,086,051	55,229
	501,528	2,925	590,607	2,907	661,744	2,700
4 (2.34 kg/person/day)	111,189	105	152,814	109	166,977	107
	13,601,628	24,126	16,972,554	28,235	21,377,002	33,187
	4,887,032	8,712	5,351,594	8,946	5,805,911	8,878
The National Solid Waste Management Strategy plan covers only 77.5 percent of Azerbaijan, excluding occupied territory (20 percent) and the Greater Baku Area (2.5 percent). The total amount generated in these areas is 964,427 tonnes/year; value is calculated from the amount generated in these areas (964,427 tonnes/year) and in the Baku area (1,965,922 tonnes/year).	2,900,944	9,725	3,329,963	10,680	3,617,967	11,039

(Table continues on next page)

Country or economy	Region	Income	Original year reported			Source
			MSW generation	Population	Year	
Bahamas, The	LAC	HIC	264,000	386,838	2015	SIDS DOCK 2015, 10
Bahrain	MENA	HIC	951,943	1,425,171	2016	Idrees and McDonnell 2016
Bangladesh	SAR	LMIC	14,778,497	155,727,053	2012	BMDF 2012
Barbados	LAC	HIC	174,815	280,601	2011	Burnside 2014
Belarus	ECA	UMIC	4,280,000	9,489,616	2015	Belarus, National Statistical Committee 2017
Belgium	ECA	HIC	4,708,000	11,274,196	2015	Eurostat 2017
Belize	LAC	UMIC	101,379	359,288	2015	IDB 2015, 3
Benin	SSA	LIC	685,936	5,521,763	1993	Achankeng 2003, 11
Bermuda	NA	HIC	82,000	64,798	2012	UNSD 2016
Bhutan	SAR	LMIC	111,314	686,958	2007	Phuntsho et al. 2007
Bolivia	LAC	LMIC	2,219,052	10,724,705	2015	Bolivia, DGGIRS 2016

Comment	2016 adjusted		2030 projected		2050 projected	
	MSW generation	Population ('000s)	MSW generation	Population ('000s)	MSW generation	Population ('000s)
5; Residential waste accounts for 70 percent of the waste collected and commercial waste accounts for 30 percent. About 77 percent of this amount is generated in New Providence.	263,946	391	317,600	440	373,151	475
4 (1.83 kg/person/day)	951,943	1,425	1,423,838	2,013	1,785,605	2,327
4 (0.29 kg/person/day)	16,380,103	162,952	22,138,475	185,585	31,162,100	201,927
Includes only mixed MSW (collected curbside in bags or containers that can be picked up by hand) entering the landfill with estimated collection coverage of 90 percent; includes cardboard, coconut husks, green waste, pellets and lumber, paper, plastic, shingles, and tires; excludes C&D waste.	178,767	285	200,673	290	223,677	280
2	4,227,784	9,480	4,935,505	9,163	5,451,248	8,571
	4,759,760	11,358	5,349,712	12,002	6,164,189	12,488
1 (urban, 1.07 kg/person/day)	102,440	367	144,792	473	223,778	592
1 (Porto Novo, 0.5 kg/person/day)	1,401,386	10,872	2,166,407	15,628	4,202,189	23,930
3	102,261	62	104,677	59	100,274	53
1 (urban, 0.53 kg/person/day)	152,647	798	249,472	914	367,260	994
Includes domestic waste and waste from public areas and markets; excludes slaughterhouse and hospital waste (together totaling 144,155.6 tonnes/year); data estimated based on population from census projections and per capita generation estimates from in-country studies in sample cities and reference values in the region for cities of similar size and rural areas.	2,276,967	10,888	3,288,932	13,158	5,214,928	15,903

(Table continues on next page)

Country or economy	Region	Income	Original year reported MSW generation	Population	Year	Source
Bosnia and Herzegovina	ECA	UMIC	1,248,718	3,535,961	2015	Bosnia and Herzegovina, BHAS 2016, 1
Botswana	SSA	UMIC	210,854	2,014,866	2010	Seanama Conservation 2012; Botswana, Statistics Botswana 2011
Brazil	LAC	UMIC	79,889,010	205,962,108	2015	ABRELPE 2015, 19
British Virgin Islands	LAC	HIC	21,099	20,645	2000	Treasure n.d., 3
Brunei Darussalam	EAP	HIC	216,253	423,196	2016	Brunei, Department of Environment, Parks and Recreation 2015
Bulgaria	ECA	UMIC	3,011,000	7,177,991	2015	Eurostat 2017
Burkina Faso	SSA	LIC	2,575,251	18,110,624	2015	Cissé 2015
Burundi	SSA	LIC	1,872,016	6,741,569	2002	UNECA-UNEP-UNIDO-ARSCP 2011
Cabo Verde	SSA	LMIC	132,555	513,979	2012	de Carvalho 2013, 15
Cambodia	EAP	LMIC	1,089,000	15,270,790	2014	Modak et al. 2017, 214
Cameroon	SSA	LMIC	3,270,617	21,655,715	2013	Mbue, Bitondo, and Balgah 2015
Canada	NA	HIC	25,103,034	35,544,564	2014	Canada, Statistics Canada 2016
Cayman Islands	LAC	HIC	60,000	59,172	2014	Amec Foster Wheeler 2016
Central African Republic	SSA	LIC	1,105,983	4,515,392	2014	UN OCHA 2014; UN DESA 2014b
Chad	SSA	LIC	1,358,851	11,887,202	2010	Simos and de Leeuw 2017, 94; UN DESA 2014b
Channel Islands	ECA	HIC	178,933	164,541	2016	States of Guernsey 2017; States of Jersey 2018
Chile	LAC	HIC	7,530,879	17,910,000	2009	Chile, CONAMA 2010, 12
China	EAP	UMIC	220,402,706	1,403,500	2015	Ji et al. 2016, 2
Colombia	LAC	UMIC	13,475,241	48,653,000	2011	IDB 2012, 11 and 25
Comoros	SSA	LIC	93,134	796,000	2015	World Bank 2015a
Congo, Dem. Rep.	SSA	LIC	14,385,226	78,736,153	2016	Tshitala Kalula 2016; UN DESA 2014b
Congo, Rep.	SSA	LMIC	894,237	5,126,000	1993	Achankeng 2003, 11
Costa Rica	LAC	UMIC	1,525,982	4,857,274	2014	Costa Rica, Ministry of Health 2016, 13

Comment	2016 adjusted		2030 projected		2050 projected	
	MSW generation	Population ('000s)	MSW generation	Population ('000s)	MSW generation	Population ('000s)
3; and percentage of population using municipal services across various municipalities.	1,261,143	3,517	1,457,111	3,405	1,588,584	3,058
1 (Gaborone, 85 tonnes/ month)	252,462	2,250	363,790	2,800	516,517	3,421
4 (218,874 tonnes/day)	79,081,401	207,653	96,693,974	225,472	114,304,745	232,688
4 (2.8 kg/person/day)	41,960	31	63,605	35	75,942	38
4 (1.4 kg/person/day)	216,253	423	262,788	490	307,979	537
	3,049,324	7,131	3,306,089	6,431	3,295,494	5,424
1 (big cities, 0.7 kg/person/ day; small cities, 0.5 kg/ person/day; average used); 2	2,659,191	18,646	4,265,523	27,382	8,807,490	43,207
1 (Bujumbura, 511 kg/person/year)	2,950,090	10,524	4,228,365	15,799	8,367,259	25,762
	139,864	540	191,675	635	274,533	734
	1,159,859	15,762	1,702,523	18,798	2,641,058	22,019
1 (Douala, 0.54 kg/person/day)	3,621,758	23,439	5,862,357	32,980	11,858,301	49,817
Value represents waste disposed of from residential and nonresidential sources.	25,666,127	36,290	30,384,216	40,618	36,171,524	44,949
5	76,141	61	117,277	71	150,789	81
1 (Bangui, 750 tonnes/day)	1,107,218	4,595	1,377,932	6,124	2,366,704	8,851
1 (N'Djamena, 533 tonnes/day)	1,645,769	14,453	2,564,763	21,460	5,237,093	33,636
	178,933	165	207,125	174	235,743	181
	7,530,879	17,910	9,359,890	19,637	11,403,108	20,718
	220,402,706	1,403,500	295,035,224	1,441,182	335,791,732	1,364,457
4 (33,288 tonnes/day) Data represents 1,102 out of 1,120 municipalities.	13,475,241	48,653	16,435,975	53,134	20,091,306	54,733
	93,134	796	131,021	1,062	234,683	1,463
1 (Kinshasa, 7,000 tonnes/day)	14,385,226	78,736	21,491,194	120,443	44,389,132	197,404
1 (Brazzaville, 0.6 kg/person/ day)	894,237	5,126	1,533,286	7,319	3,193,587	11,510
4 (4,000 tonnes/day)	1,525,982	4,857	1,933,590	5,417	2,389,760	5,774

(Table continues on next page)

Country or economy	Region	Income	Original year reported			Source
			MSW generation	Population	Year	
Côte d'Ivoire	SSA	LMIC	4,440,814	20,401,331	2010	Ludington 2015; UN DESA 2014b
Croatia	ECA	UMIC	1,654,000	4,203,604	2015	Eurostat 2017
Cuba	LAC	UMIC	2,692,692	11,303,687	2007	Rebelde 2007
Curaçao	LAC	HIC	24,704	153,822	2013	PricewaterhouseCoopers Aruba 2014, 11
Cyprus	ECA	HIC	541,000	1,160,985	2015	Eurostat 2017
Czech Republic	ECA	HIC	3,337,000	10,546,059	2015	Eurostat 2017
Denmark	ECA	HIC	4,485,000	5,683,483	2015	Eurostat 2017
Djibouti	MENA	LMIC	114,997	746,221	2002	IMF 2004
Dominica	LAC	UMIC	13,176	72,400	2013	World Bank 2017a, 5
Dominican Republic	LAC	UMIC	4,063,910	10,528,394	2015	Dominican Republic, Ministry of the Environment and Natural Resources 2017, 16
Ecuador	LAC	UMIC	5,297,211	16,144,368	2015	Ecuador, Ministry of Environment 2018
Egypt, Arab Rep.	MENA	LMIC	21,000,000	87,813,257	2012	GIZ and SWEEP-Net 2014a, 10
El Salvador	LAC	LMIC	1,648,996	6,164,626	2010	IDB-AIDIS-PAHO 2011, 104
Equatorial Guinea	SSA	UMIC	198,443	1,221,490	2016	Calculated (See box 1.1)
Eritrea	SSA	LIC	726,957	4,474,690	2011	Calculated (See box 1.1)
Estonia	ECA	HIC	473,000	1,315,407	2015	Eurostat 2017
Eswatini	SSA	LMIC	218,199	1,343,098	2016	Calculated (See box 2.1)

Comment	2016 adjusted		2030 projected		2050 projected	
	MSW generation	Population ('000s)	MSW generation	Population ('000s)	MSW generation	Population ('000s)
1 (Abidjan, 290 kg/person/year)	5,525,029	23,696	9,817,371	33,337	22,186,836	51,375
	1,684,219	4,213	1,703,139	3,896	1,670,840	3,461
1 (Havana, 0.7 kg/person/day; rural, 0.5 kg/person/day); excludes bulky, industrial, and medical waste; likely also excludes commercial waste, but this cannot be confirmed.	2,818,053	11,476	3,253,115	11,496	3,647,101	10,823
4 (0.44 kg/person/day)	31,787	159	45,230	172	53,398	181
	551,614	1,170	624,277	1,282	715,657	1,383
	3,389,662	10,611	3,848,146	10,528	4,245,312	10,054
	4,527,726	5,712	4,982,841	6,025	5,640,297	6,314
1 (Djibouti City, 240 kg/day)	152,359	942	217,297	1,133	332,342	1,308
5; total waste (urban and rural) collected, including household (67 percent), commercial (17 percent), institutional (5 percent), industrial (6 percent), and other (6 percent), is 12,385 tonnes per year; uncollected waste was included by using the collection coverage (94 percent).	13,542	74	17,555	78	20,671	77
4 (11,134 tonnes/day)	4,202,756	10,649	5,412,538	12,098	6,905,740	13,265
2; 9 (12,829.41 tonnes/day collected and a collection rate of 88.5 percent)	5,307,241	16,385	7,157,795	19,555	10,225,146	22,968
	23,366,729	95,689	34,213,851	119,746	55,163,107	153,433
1 (urban, 0.89 kg/person/day)	1,786,871	6,345	2,162,742	6,786	2,771,792	6,997
6	198,443	1,221	319,272	1,871	557,175	2,845
6	774,249	4,955	1,084,661	6,718	1,991,475	9,607
	475,808	1,312	523,237	1,254	553,719	1,145
6	218,199	1,343	276,577	1,666	407,836	2,081

(Table continues on next page)

Country or economy	Region	Income	Original year reported			Source
			MSW generation	Population	Year	
Ethiopia	SSA	LIC	6,532,787	99,873,033	2015	GIZ 2015 (Awadai, Bale Robe, Burie Town, Dilla, Dire Dawa, Jijiga, Jimma, Waliso, Wolkite, Wuqro); Artelia Ville et Transport 2014 (Addis Ababa); WaterAid 2015a (Adigrat, Axum, Bahir Dar, Bule Hora, Yirgachefe, Injibara, Finoteselam); WaterAid 2015b (Ambo, Hosanna, Bishoftu, Fitche, Gerbe Guracha, Holeta, Yirgalem); Anon 2015 (Adola Woyu, Weldiya, Tepi, Maichew, Halaba Kulito Town, Dembi Dolo, Debre Tabor, Bati)
Faroe Islands	ECA	HIC	61,000	48,842	2014	Nordic Competition Authorities 2016, 57
Fiji	EAP	UMIC	189,390	867,086	2011	Fiji, Department of Environment 2011, 13
Finland	ECA	HIC	2,738,000	5,479,531	2015	Eurostat 2017
France	ECA	HIC	33,399,000	66,624,068	2015	Eurostat 2017
French Polynesia	EAP	HIC	147,000	273,528	2013	French Polynesia, DIREN 2017, 223
Gabon	SSA	UMIC	238,102	1,086,137	1995	Mombo and Edou 2005, 90
Gambia, The	SSA	LIC	193,441	1,311,349	2002	Sanneh et al. 2011, 3; UN DESA 2014b
Georgia	ECA	LMIC	800,000	3,717,100	2015	Particip 2015, 8
Germany	ECA	HIC	51,046,000	81,686,611	2015	Eurostat 2017
Ghana	SSA	LMIC	3,538,275	21,542,009	2005	Puopiel 2010, 21
Gibraltar	ECA	HIC	16,954	33,623	2012	Gibraltar, Ministry for the Environment 2017, 30
Greece	ECA	HIC	5,477,424	10,892,413	2014	Greece, ELSTAT 2017
Greenland	ECA	HIC	50,000	56,905	2010	Eisted and Christensen 2011, 1
Grenada	LAC	UMIC	29,536	105,481	2012	Caribbean Development Bank 2014

Comment	2016 adjusted		2030 projected		2050 projected	
	MSW generation	Population ('000s)	MSW generation	Population ('000s)	MSW generation	Population ('000s)
1 (various cities, 0.30 kg/person/day)	6,727,941	102,403	10,040,763	139,620	18,102,122	190,870
	65,882	49	72,356	52	83,920	55
1 (urban, 0.78 kg/person/day; rural, 0.4 kg/person/day)	206,277	899	242,350	970	298,039	998
	2,769,576	5,503	3,079,571	5,739	3,449,266	5,866
	32,544,914	64,721	36,021,363	67,894	40,862,922	70,609
5	139,585	280	174,067	307	199,138	326
1 (Libreville, 0.685 kg/person/day)	403,931	1,980	578,036	2,594	924,679	3,516
1 (Banjul, 0.54 kg/person/day)	301,751	2,039	503,966	3,001	1,078,463	4,562
3	854,577	3,925	998,425	3,748	1,136,220	3,394
	51,410,863	81,915	54,399,513	82,187	57,050,957	79,238
4 (0.45 kg/person/day); 5	5,287,958	28,207	8,142,202	37,294	14,272,518	51,270
Value represents MSW, excluding mattresses and nonhazardous bulky waste.	18,761	34	20,279	36	22,973	37
	5,636,374	11,184	5,966,360	10,784	6,379,219	9,982
5	53,601	56	56,336	57	58,128	54
Value includes HH (20,818 tonnes/year), institutional (1,017 tonnes/year), and commercial (5,560 tonnes/year) waste disposed of in Perserverance Landfill and all the waste in Dumfries Landfill (1,639 tonnes/year) totaling 29,034 tonnes/year. This value is then adjusted by the collection rate of 98.3 percent; C&D, shipping, industrial, and green waste excluded.	32,359	107	37,194	112	43,325	110

(Table continues on next page)

| Country or economy | Region | Income | Original year reported | | | Source |
			MSW generation	Population	Year	
Guam	EAP	HIC	141,500	159,973	2012	Guam 2013, 10
Guatemala	LAC	LMIC	2,756,741	16,252,429	2015	IDB 2015, 3
Guinea	SSA	LIC	596,911	8,132,552	1996	Barry 2002; UN DESA 2014b
Guinea-Bissau	SSA	LIC	289,514	1,770,526	2015	Ferrari et al. 2016, 2
Guyana	LAC	UMIC	179,252	746,556	2010	Guyana, Ministry of Communities n.d., 11 (Table 2)
Haiti	LAC	LIC	2,309,852	10,847,334	2015	SWANA Haiti Response Team 2010, 4; Naquin 2016, 12
Honduras	LAC	LMIC	2,162,028	9,112,867	2016	Honduras, DGA 2017
Hong Kong SAR, China	EAP	HIC	5,679,816	7,305,700	2015	Hong Kong, Environmental Protection Department 2016
Hungary	ECA	HIC	3,712,000	9,843,028	2015	Eurostat 2017
Iceland	ECA	HIC	525,000	330,815	2015	Iceland, Statistics Iceland 2015, 429

Comment	2016 adjusted		2030 projected		2050 projected	
	MSW generation	Population ('000s)	MSW generation	Population ('000s)	MSW generation	Population ('000s)
Reported as a range of 129,000–154,000 tonnes/year (average used).	134,551	163	170,308	181	195,298	193
1 (urban, 0.61 kg/person/day)	2,824,598	16,582	3,990,278	21,203	6,307,100	26,968
1 (Conakry, 0.31 kg/person/day)	941,169	12,396	1,757,060	17,631	4,102,204	26,852
1 (Bissau, 0.6 kg/person/day)	297,640	1,816	441,963	2,493	894,814	3,603
4 (491.1 tonnes/day); includes HH and commercial waste; 491.1 tonnes/day is a weighted average of waste generation in 10 regions; per capita waste generation (0.73 kg/person/day) was measured for the most populous region, Region 4, while remainder used assumptions based on reference values.	202,463	773	244,517	825	293,510	822
1 (Port au Prince metro area, 0.7 kg/person/day; rural, 0.41 kg/person/day); rural rate is for Cap-Haïtien data, which is representative of the rest of the country and involves both rural and urban populations; Port au Prince data are from 2010 and Cap-Haïtien's from 2016.	2,309,852	10,847	2,975,484	12,544	4,693,120	14,041
4 (0.65 kg/person/day); 7; based on 62 percent of the population in Honduras.	2,162,028	9,113	3,050,449	11,147	4,787,863	13,249
4 (2.13 kg/person/day); 1.39 kg/person/day is the rate of 65 percent of MSW disposed of at landfill; when 35 percent of recovered MSW is factored in, value increases to 2.13 kg/person/day.	5,710,414	7,303	6,858,836	7,987	7,637,326	8,253
	3,715,742	9,753	3,885,730	9,235	3,989,253	8,279
5	539,686	332	637,438	366	755,434	390

(Table continues on next page)

Country or economy	Region	Income	Original year reported			Source
			MSW generation	Population	Year	
India	SAR	LMIC	168,403,240	1,071,477,855	2001	UNICEF-India, Ministry of Rural Development 2008; India, Ministry of Home Affairs 2001; Kumar et al. 2009
Indonesia	EAP	LMIC	65,200,000	261,115,456	2016	Indonesia, Ministry of Environment and Forestry and Ministry of Industry 2016, 4
Iran, Islamic Rep.	MENA	UMIC	17,885,000	80,277,428	2017	Abedini 2017
Iraq	MENA	UMIC	13,140,000	36,115,649	2015	Iraq, Ministry of Environment 2015
Ireland	ECA	HIC	2,692,537	4,586,897	2012	Ireland, EPA 2014, 1
Isle of Man	ECA	HIC	50,551	80,759	2011	Isle of Man, Department of Infrastructure n.d., 10
Israel	MENA	HIC	5,400,000	8,380,100	2015	Israel, Ministry of Environmental Protection 2016
Italy	ECA	HIC	29,524,000	60,730,582	2015	Eurostat 2017
Jamaica	LAC	UMIC	1,051,695	2,881,355	2016	Jamaica, NSWMA n.d., 7

Comment	2016 adjusted		2030 projected		2050 projected	
	MSW generation	Population ('000s)	MSW generation	Population ('000s)	MSW generation	Population ('000s)
Calculated based on daily per capita waste generation rates published by the Central Public Health and Environmental Engineering Organizations, segmented by population groups. Population was taken from 2011 census. Calculations as follows: Population > 5 million (85,188,627), 0.5 kg/person/day; population between 2 million and 5 million (28,850,634), 0.35 kg/person/day; population between 1 million and 2 million (46,686,245), 0.27 kg/person/day; population between 500,000 and 1 million (30,235,593), 0.25 kg/person/day; population between 100,000 and 500,000 (73,930,414), 0.21 kg/person/day; for towns and villages > 100,000 population (833,748,852), assumed that the waste generation rate is half that of the lowest population group (0.11 kg/person/day); value is likely a conservative estimate.	277,136,133	1,324,171	387,770,524	1,512,985	543,277,457	1,658,978
	65,200,000	261,115	87,958,248	295,595	118,551,290	321,551
2; estimated based on reports from some cities and some rural areas.	17,653,936	80,277	21,303,899	88,863	25,597,026	93,553
4 (1.2 kg/person/day)	13,967,851	37,203	21,053,906	53,298	34,328,393	81,490
	3,157,225	4,726	3,692,571	5,220	4,322,409	5,801
7	56,476	84	63,371	91	74,679	97
Value represents municipal and commercial waste.	5,322,248	8,192	7,108,848	9,984	10,038,606	12,577
	29,009,742	59,430	29,855,267	58,110	30,839,601	55,093
4 (1 kg/person/day); per capita generation calculated by a collection sampling exercise in six urban and rural communities.	1,051,695	2,881	1,156,300	2,933	1,271,212	2,704

(Table continues on next page)

Country or economy	Region	Income	Original year reported			Source
			MSW generation	Population	Year	
Japan	EAP	HIC	43,981,000	127,141,000	2015	Japan, Ministry of the Environment 2015
Jordan	MENA	LMIC	2,529,997	8,413,464	2013	Al-Jayyousi 2015, 22
Kazakhstan	ECA	UMIC	4,659,740	16,791,425	2012	World Bank n.d., 42
Kenya	SSA	LMIC	5,595,099	41,350,152	2010	Okot-Okumu 2012, 4
Kiribati	EAP	LMIC	35,724	114,395	2016	SPREP 2016, 21
Korea, Rep.	EAP	HIC	18,218,975	50,746,659	2014	Kho and Lee 2016, 23
Kosovo	ECA	LMIC	319,000	1,801,800	2015	Eurostat 2017
Kuwait	MENA	HIC	1,750,000	2,998,083	2010	Kuwait, Partnerships Technical Bureau 2014, 4
Kyrgyz Republic	ECA	LMIC	1,113,300	5,956,900	2015	Kyrgyzstan, NSC 2016, 62 (Table 5.9)
Lao PDR	EAP	LMIC	351,900	6,663,967	2015	Keohanam 2017
Latvia	ECA	HIC	857,000	1,977,527	2015	Eurostat 2017
Lebanon	MENA	UMIC	2,040,000	5,603,279	2014	GIZ and SWEEP-Net 2014b, 8; UNDP 2014
Lesotho	SSA	LMIC	73,457	1,965,662	2006	Lesotho, Bureau of Statistics 2013; Lesotho, Bureau of Statistics 2006
Liberia	SSA	LIC	564,467	3,512,932	2007	UNEP 2007; UN DESA 2014b
Libya	MENA	UMIC	2,147,596	6,193,501	2011	Omran, Altawati, and Davis 2017, 5
Liechtenstein	ECA	HIC	32,382	36,545	2015	Liechtenstein, Office of Statistics 2018, 7
Lithuania	ECA	HIC	1,300,000	2,904,910	2015	Eurostat 2017
Luxembourg	ECA	HIC	356,000	569,604	2015	Eurostat 2017
Macao SAR, China	EAP	HIC	377,942	612,167	2016	Macao SAR, China, DSEC 2017
Macedonia, FYR	ECA	UMIC	796,585	2,081,206	2016	Macedonia, MAKStat 2017
Madagascar	SSA	LIC	3,768,759	24,894,551	2016	World Bank 2016a, 5
Malawi	SSA	LIC	1,297,844	16,577,147	2013	Barré 2014; UN DESA 2014a

Comment	2016 adjusted		2030 projected		2050 projected	
	MSW generation	Population ('000s)	MSW generation	Population ('000s)	MSW generation	Population ('000s)
Excludes disaster waste.	44,374,189	127,749	45,019,046	121,581	43,315,197	108,794
	2,793,380	9,456	3,825,435	11,122	6,351,694	14,188
9 (3,588,000 tonnes/year collected and a collection rate of 77 percent)	5,126,019	17,988	6,850,097	20,301	8,512,123	22,959
1 (Nairobi, 0.6 kg/ person/day)	6,844,079	48,462	10,513,071	66,960	19,033,007	95,467
8	35,724	114	70,876	142	115,089	178
4 (49,915 tonnes/day)	18,576,898	50,792	22,435,453	52,702	24,624,834	50,457
	323,281	1,802	484,974	1,802	645,955	1,802
	2,290,389	4,053	2,894,529	4,874	3,613,973	5,644
	1,120,523	5,956	1,566,360	6,997	2,475,253	8,113
2	364,463	6,758	522,053	8,049	748,378	9,163
	864,936	1,971	881,848	1,747	861,239	1,517
UNDP 2014 estimates that the incremental daily quantity of MSW attributed to refugees is expected to reach 324,568 tonnes/year by 2014; this value is significant and is equivalent to about 15.7 percent of the waste generated by Lebanese residents before the crisis.	2,148,803	6,007	2,302,862	5,369	2,862,432	5,412
1 (Maseru City, 60 kg/ person/year)	87,981	2,204	117,518	2,608	193,270	3,203
1 (Monrovia, 780 tonnes/ day)	722,949	4,614	988,354	6,495	1,910,290	9,804
4 (0.95 kg/person/day)	2,419,759	6,293	3,631,710	7,342	4,617,447	8,124
Value represents total urban waste only.	35,486	38	39,939	41	46,168	43
	1,320,616	2,908	1,382,158	2,718	1,363,525	2,407
	360,964	576	433,768	675	524,875	796
Value includes domestic waste and waste produced by businesses.	377,942	612	481,342	746	575,184	876
	796,585	2,081	933,411	2,076	1,056,395	1,931
1 (Antananarivo, 0.61 kg/ person/day); 5	3,768,759	24,895	5,587,354	35,592	10,522,518	53,803
1 (Lilongwe and Blantyre, 0.37 kg/person/day)	1,415,204	18,092	2,117,841	26,578	4,081,844	41,705

(Table continues on next page)

Country or economy	Region	Income	Original year reported			Source
			MSW generation	Population	Year	
Malaysia	EAP	UMIC	12,982,685	30,228,017	2014	UNCRD and IGES 2017, xii
Maldives	SAR	UMIC	211,506	409,163	2015	Maldives, WMPDC and MEE 2017
Mali	SSA	LIC	1,937,354	16,006,670	2012	World Bank 2014, 66
Malta	MENA	HIC	269,000	431,874	2015	Eurostat 2017
Marshall Islands	EAP	UMIC	8,614	52,793	2013	Pattle Delamore Partners 2015, 17
Mauritania	SSA	LMIC	454,000	3,506,288	2009	GIZ and SWEEP-Net 2010a
Mauritius	SSA	UMIC	438,000	1,263,473	2016	Mauritius, Ministry of Social Security, National Solidarity, and Environment and Sustainable Development 2017
Mexico	LAC	UMIC	53,100,000	125,890,949	2015	Mexico, SEMARNAT 2016, 434
Micronesia, Fed. Sts.	EAP	LMIC	26,040	104,937	2016	SPREP 2016, 21
Moldova	ECA	LMIC	3,981,200	3,554,108	2015	Moldova, Statistica Moldovei 2016
Monaco	ECA	HIC	46,000	37,783	2012	UNSD 2016
Mongolia	EAP	LMIC	2,900,000	3,027,398	2016	Delgerbayar 2016, 4
Montenegro	ECA	UMIC	332,000	622,159	2015	Eurostat 2017
Morocco	MENA	LMIC	6,852,000	34,318,082	2014	GIZ and SWEEP-Net 2014c, 7
Mozambique	SSA	LIC	2,500,000	27,212,382	2014	Tas and Belon 2014, 9
Myanmar	EAP	LMIC	4,677,307	46,095,462	2000	Thein 2010, 6
Namibia	SSA	UMIC	256,729	1,559,983	1993	Achankeng 2003, 11
Nauru	EAP	UMIC	6,192	13,049	2016	SPREP 2016, 21
Nepal	SAR	LIC	1,768,977	28,982,771	2016	Nepal, SWMTSC 2017; ADB 2013
Netherlands	ECA	HIC	8,855,000	16,939,923	2015	Eurostat 2017

Comment	2016 adjusted		2030 projected		2050 projected	
	MSW generation	Population ('000s)	MSW generation	Population ('000s)	MSW generation	Population ('000s)
	13,723,342	31,187	18,235,817	36,815	23,733,545	41,729
2; based on Maldives Ministry of Environment and Energy data for households in Male (1.7 kg/person/day), other islands (0.8 kg/person/day), and resorts, hotels, and guest houses (3.5 kg/bed/day).	224,663	428	300,525	512	393,328	576
1 (Sikasso, 46,770 tonnes/year)	2,207,589	17,995	3,515,355	27,057	7,084,361	44,020
	270,442	429	303,995	440	324,623	419
4 (23.6 tonnes/day)	8666	53	14,195	56	20,046	66
	572,992	4,301	919,925	6,077	1,771,918	8,965
4 (1,200 tonnes/day)	437,535	1,262	518,359	1,287	571,593	1,221
	54,151,287	127,540	69,638,974	147,540	90,440,574	164,279
8	129,821	523	207,574	589	298,646	656
5; legislation does not clearly differentiate between industrial and municipal waste. Existing law defines waste from production and consumption; hence, waste statistics include both types of waste.	4,622,874	4,060	5,636,646	3,844	6,588,017	3,293
3	50,685	38	56,417	41	68,391	46
5	2,900,000	3,027	4,337,475	3,561	6,295,598	4,075
	339,542	629	368,880	625	399,240	588
	7,126,270	35,277	10,160,132	40,874	15,157,504	45,660
	2,644,873	28,829	4,124,044	42,439	8,750,664	67,775
4 (0.278 kg/person/day)	7,451,835	52,885	9,315,917	58,916	11,207,310	62,359
1 (Windhoek, 0.7 kg/person/day)	501,797	2,480	738,810	3,246	1,205,787	4,339
8; Nauru is 100 percent urban according to source	5,384	11	5,200	11	6,139	11
60 additional municipalities were newly formed recently; the two sources provide waste generation for 58 cities and the additional 60 cities, respectively; average waste generated per day for 118 cities is estimated to be 1,854 tonnes/day.	1,768,977	28,983	2,205,525	33,168	2,968,223	36,107
	8,936,530	16,987	9,816,231	17,594	10,677,957	17,518

(Table continues on next page)

Country or economy	Region	Income	Original year reported			Source
			MSW generation	Population	Year	
New Caledonia	EAP	HIC	108,157	278,000	2016	SPREP 2016
New Zealand	EAP	HIC	3,405,000	4,692,700	2016	OECD 2018
Nicaragua	LAC	LMIC	1,528,816	5,737,723	2010	IDB-AIDIS-PAHO 2011
Niger	SSA	LIC	1,865,646	8,842,415	1993	Achankeng 2003, 11
Nigeria	SSA	LMIC	27,614,830	154,402,181	2009	Oguntoyinbo 2012, 1
Northern Mariana Islands	EAP	HIC	32,761	54,036	2013	Mohee et al. 2015
Norway	ECA	HIC	2,187,000	5,188,607	2015	Eurostat 2017
Oman	MENA	HIC	1,734,885	3,960,925	2014	Be'ah 2016, 13
Pakistan	SAR	LMIC	30,760,000	193,203,476	2017	Korai, Mahar, and Uqaili 2017
Palau	EAP	HIC	9,427	21,503	2016	SPREP 2016, 21
Panama	LAC	UMIC	1,472,262	3,969,249	2015	IDB 2015, 3
Papua New Guinea	EAP	LMIC	1,000,000	7,755,785	2014	ADB 2014a, 1
Paraguay	LAC	UMIC	1,818,501	6,639,119	2015	IDB-AIDIS-PAHO 2011, 104
Peru	LAC	UMIC	8,356,711	30,973,354	2014	Peru, Ministry of Environment 2016, 20; Peru, Ministry of Environment 2014, 33
Philippines	EAP	LMIC	14,631,923	103,320,222	2016	Philippines, NSWMC 2017
Poland	ECA	HIC	10,863,000	37,986,412	2015	Eurostat 2017
Portugal	ECA	HIC	4,710,000	10,401,062	2014	Eurostat 2017
Puerto Rico	LAC	HIC	4,170,953	3,473,181	2015	Energy Answers 2012, 41
Qatar	MENA	HIC	1,000,990	2,109,568	2012	Qatar, MDPS 2014, 92
Romania	ECA	UMIC	4,895,000	19,815,481	2015	Eurostat 2017

Comment	2016 adjusted		2030 projected		2050 projected	
	MSW generation	Population ('000s)	MSW generation	Population ('000s)	MSW generation	Population ('000s)
8	106,086	273	132,841	321	168,274	378
	3,381,877	4,661	3,971,657	5,213	4,789,174	5,711
4 (0.73 kg/person/day); 7	1,787,370	6,150	2,363,847	7,046	3,502,392	7,876
1 (Niamey, 1 kg/person/day)	4,281,415	20,673	7,164,740	34,994	16,015,498	68,454
4 (0.49 kg/person/day)	34,572,968	185,990	54,806,190	264,068	107,077,289	410,638
1 (Saipan, 1.81 kg/person/day; average regional rural, 0.5 kg/person/day)	30,922	55	36,345	57	36,190	52
	2,216,799	5,255	2,593,368	5,959	3,070,182	6,802
4 (1.2 kg/person/day)	1,928,958	4,425	2,710,244	5,897	3,385,564	6,757
	30,352,981	193,203	42,427,624	244,248	66,377,808	306,940
8	9,427	22	19,117	25	22,944	28
1 (urban, 1.22 kg/person/day)	1,516,612	4,034	2,194,682	4,884	3,074,132	5,827
Reported as more than 1 million tonnes/year.	1,052,408	8,085	1,595,910	10,487	2,844,877	13,871
1 (urban, 0.94 kg/person/day)	1,862,514	6,725	2,484,878	7,845	3,595,736	8,897
Calculated from urban (7,497,482 tonnes/year) and rural (859,229.13 tonnes/year) waste generation; urban value reported in Plan Nacional as 64 percent (4,798,388 tonnes/year) generated by HH and 26 percent (1,949,345 tonnes/year) by non-HH sources; rural value is calculated by World Bank team using an estimate of 0.35 kg/person/day based on reporting for all urban districts multiplied by the rural population in Peru in 2014.	8,737,853	31,774	12,466,705	36,807	17,441,927	41,620
4 (40,087.46 tonnes/day)	14,631,923	103,320	20,039,044	125,372	29,275,773	151,293
	11,059,953	38,224	12,000,866	36,616	11,941,493	32,390
	4,776,650	10,372	4,890,090	9,877	4,941,153	8,995
Value is a projection based on historical waste generation.	4,392,515	3,668	4,607,101	3,593	4,619,340	3,282
4 (1.30 kg/person/day); if bulky waste and tires are included, the per capita MSW generation rate increases to 3.79 kg/person/day.	1,195,225	2,570	1,592,401	3,232	1,864,992	3,773
	4,993,965	19,778	5,301,420	18,464	5,308,936	16,397

(Table continues on next page)

Country or economy	Region	Income	Original year reported			Source
			MSW generation	Population	Year	
Russian Federation	ECA	UMIC	60,000,000	143,201,676	2012	Tekes 2013, 11
Rwanda	SSA	LIC	4,384,969	11,917,508	2016	Isugi and Niu 2016
Saint Martin (French part)	LAC	HIC	15,480	30,959	2012	Sterviinou n.d.
Samoa	EAP	UMIC	27,399	187,665	2011	SPREP 2016, 21
San Marino	ECA	HIC	17,175	33,203	2016	San Marino, AASS 2016
São Tomé and Príncipe	SSA	LMIC	25,587	191,266	2014	Dias, Vaz, and Carvalho 2014
Saudi Arabia	MENA	HIC	16,125,701	31,557,144	2015	Nizami 2015
Senegal	SSA	LIC	2,454,059	15,411,614	2016	Senegal, UCG 2016, 23
Serbia	ECA	UMIC	1,840,000	7,095,383	2015	Eurostat 2017
Seychelles	SSA	HIC	48,000	88,303	2012	Talma and Martin 2013, 5
Sierra Leone	SSA	LIC	610,222	5,439,695	2004	Gogra et al. 2010, 2
Singapore	EAP	HIC	7,704,300	5,607,283	2017	Singapore, NEA 2017
Slovak Republic	ECA	HIC	1,784,000	5,423,801	2015	Eurostat 2017
Slovenia	ECA	HIC	926,000	2,063,531	2015	Eurostat 2017
Solomon Islands	EAP	LMIC	179,972	563,513	2013	ADB 2014b, 1
Somalia	SSA	LIC	2,326,099	14,317,996	2016	Calculated (See box 1.1)
South Africa	SSA	UMIC	18,457,232	51,729,345	2011	South Africa, Department of Environmental Affairs 2012
South Sudan	SSA	LIC	2,680,681	11,177,490	2013	UNEP 2013, 18
Spain	ECA	HIC	20,151,000	46,447,697	2015	Eurostat 2017
Sri Lanka	SAR	LMIC	2,631,650	21,203,000	2016	Sri Lanka, Ministry of Mahaweli Development and Environment 2016, v
St. Kitts and Nevis	LAC	HIC	32,892	54,288	2015	SIDS DOCK 2015, 15

Comment	2016 adjusted		2030 projected		2050 projected	
	MSW generation	Population ('000s)	MSW generation	Population ('000s)	MSW generation	Population ('000s)
	59,585,899	143,965	67,001,631	140,543	71,574,530	132,731
1 (Kigali, reported as a range of 1.8–2.0 kg/person/day; average used)	4,384,969	11,918	6,555,912	16,024	11,586,425	21,886
4 (500 kg/person/year)	19,322	31	25,450	31	28,535	31
4 (0.4 kg/person/day); 7	28,964	195	35,111	212	49,216	243
	17,175	33	17,018	35	18,686	35
4 (70.1 tonnes/day)	26,999	200	35,319	268	64,173	380
4 (1.4 kg/person/day)	16,455,464	32,276	20,986,707	39,480	25,183,676	45,056
4 (6,723.45 tonnes/day)	2,454,059	15,412	3,957,017	22,123	8,059,355	34,031
A new methodology to collect MSW generation data was introduced in 2010, which requires public utility companies to report collected waste amounts and MSW composition. In 2013, data were delivered by 106 of 168 companies; data reported by some companies are still based on estimates.	2,319,171	8,820	2,408,682	8,355	2,392,222	7,447
5	53,921	94	58,271	98	72,626	97
1 (Freetown, 0.45 kg/person/ day)	829,206	7,396	1,157,579	9,720	1,998,055	12,972
5	7,629,509	5,622	9,284,685	6,342	9,989,340	6,575
	1,813,640	5,444	2,024,455	5,387	2,132,309	4,965
	943,902	2,078	1,029,557	2,059	1,090,649	1,942
4 (reported a range of 0.75–1.0 kg/person/day, average used)	192,172	599	291,573	773	535,497	1,033
6	2,326,099	14,318	3,411,381	21,535	7,291,620	35,852
Includes municipal, commercial, and industrial waste; excludes C&D, hazardous, and inert waste.	20,102,994	56,015	27,094,596	64,466	36,766,292	72,755
1 (Juba, 1.11 kg/person/day)	2,854,926	12,231	3,989,661	17,254	7,530,449	25,366
	20,361,483	46,348	21,226,169	46,115	21,829,247	44,395
4 (7,210 tonnes/day). The data are for 22 out of 25 districts in Sri Lanka and exclude approximately one-quarter of the population.	2,581,444	20,798	3,168,447	21,475	3,746,891	20,792
4 (St. Kitts: 2.08 kg/person/ day and Nevis: 1.52 kg/ person/day)	33,380	55	59,629	61	69,926	63

(Table continues on next page)

Country or economy	Region	Income	Original year reported			Source
			MSW generation	Population	Year	
St. Lucia	LAC	UMIC	77,616	177,206	2015	St. Lucia, SLSWMA 2015, 47
St. Vincent and the Grenadines	LAC	UMIC	31,561	109,455	2015	SIDS DOCK 2015, 15
Sudan	SSA	LMIC	2,831,291	38,647,803	2015	Elbaroudi, Ahmed, and Adam 2015, 9
Suriname	LAC	UMIC	78,620	526,103	2010	IDB 2010; Zuilen 2006
Sweden	ECA	HIC	4,377,000	9,799,186	2015	Eurostat 2017
Switzerland	ECA	HIC	6,056,000	8,372,098	2016	OECD 2018
Syrian Arab Republic	MENA	LMIC	4,500,000	20,824,893	2009	GIZ and SWEEP-Net 2010b, 5
Taiwan, China	EAP	HIC	7,336,000	23,434,000	2015	Chen 2016
Tajikistan	ECA	LMIC	1,787,400	8,177,809	2013	Tajikistan, Tajstat 2017
Tanzania	SSA	LIC	9,276,995	49,082,997	2012	Tanzania, NBS and OCGS 2014
Thailand	EAP	UMIC	26,853,366	68,657,600	2015	Thailand, PCD 2015, 74
Timor-Leste	EAP	LMIC	63,875	1,268,671	2016	Timor-Leste, Ministry of Commerce, Industry and the Environment 2016, 1
Togo	SSA	LIC	1,109,030	7,228,915	2014	CCAC n.d.
Tonga	EAP	UMIC	17,238	104,951	2012	ADB 2014c

Comment	2016 adjusted		2030 projected		2050 projected	
	MSW generation	Population ('000s)	MSW generation	Population ('000s)	MSW generation	Population ('000s)
4 (1.2 kg/person/day); Deglos Sanitary Landfill received 51,661 tonnes and the Vieux Fort Solid Waste Management Facility received 20,228 tonnes. The total generation was accounted for by dividing by the collection efficiency.	78,361	178	91,811	186	106,930	182
4 (0.79 kg/person/day)	31,761	110	39,210	112	45,567	109
1 (Khartoum, 0.2–0.4 kg/person/day; average used)	2,922,225	39,579	4,492,595	54,842	8,214,056	80,386
1 (urban, 0.47 kg/person/day, based on Paramaribo data; rural, 0.29 kg/person/day, based on data from other districts); 7	82,609	558	104,605	617	133,249	648
	4,426,933	9,838	5,122,838	10,712	6,019,418	11,626
	6,077,441	8,402	6,945,435	9,204	8,039,954	9,880
	3,849,718	18,430	6,594,549	26,608	11,170,733	34,021
28 percent of total waste (26,200,000 tonnes/year) comprises MSW; remaining is IW.	6,884,963	23,557	8,040,360	24,151	8,168,078	22,771
	1,968,475	8,735	3,091,105	11,194	5,633,844	14,521
Calculated based on summing amount of waste collected by company or authority, burned, dumped on roadside, buried, and other (bush, open space).	10,860,140	55,572	18,545,453	83,702	39,824,577	138,082
	27,268,302	68,864	32,484,794	69,626	37,342,182	65,372
4 (175 tonnes/day)	63,875	1,269	91,347	1,704	161,765	2,421
1 (Lome, 220 kg/person/year)	1,169,455	7,606	1,702,085	10,507	3,083,704	15,298
4 (0.5 kg/person/day); 7	17,849	107	27,763	121	38,277	140

(Table continues on next page)

Country or economy	Region	Income	Original year reported			Source
			MSW generation	Population	Year	
Trinidad and Tobago	LAC	HIC	727,874	1,328,100	2010	Trinidad and Tobago, EMA n.d., 54
Tunisia	MENA	LMIC	2,700,000	11,143,908	2014	Brahim 2017, 2
Turkey	ECA	UMIC	31,283,000	78,271,472	2015	Eurostat 2017
Turkmenistan	ECA	UMIC	500,000	5,366,277	2013	Zoï Environment Network 2013, 25
Tuvalu	EAP	UMIC	3,989	11,097	2016	SPREP 2016, 21
Uganda	SSA	LIC	7,045,050	35,093,648	2011	Okot-Okumu and Nyenje 2011
Ukraine	ECA	LMIC	15,242,025	45,004,645	2016	Ukraine, SSC 2017
United Arab Emirates	MENA	HIC	5,413,453	9,269,612	2016	Idrees and McDonnell 2016
United Kingdom	ECA	HIC	31,567,000	65,128,861	2015	Eurostat 2017
United States	NA	HIC	258,000,000	318,563,456	2014	U.S. EPA 2014, 2
Uruguay	LAC	HIC	1,260,140	3,431,552	2015	IDB 2015, 3
Uzbekistan	ECA	LMIC	4,000,000	29,774,500	2012	ADB 2012, 1
Vanuatu	EAP	LMIC	70,225	270,402	2016	SPREP 2016, 21
Venezuela, RB	LAC	UMIC	9,779,093	29,893,080	2012	Venezuela, INE 2013, 5
Vietnam	EAP	LMIC	9,570,300	86,932,500	2010	Nguyen, Heaven, and Banks 2014, 366
Virgin Islands (U.S.)	LAC	HIC	146,500	105,784	2011	Davis, Haase, and Warren 2011, 9

Comment	2016 adjusted		2030 projected		2050 projected	
	MSW generation	Population ('000s)	MSW generation	Population ('000s)	MSW generation	Population ('000s)
Value is an extrapolated number from 2009 (650,000 tonnes), which used estimates of waste going into major landfills and assumed an amount for the rest of the landfills. "Of this figure, about one third of the waste was generated from ICI sources whilst the majority of two thirds from household sources. Based on Trinidad's population, it is further estimated that 0.54 tonnes of waste is generated per capita per year amounting to 1.50 kilograms per person per day."	731,213	1,365	805,080	1,374	848,091	1,295
Reported as about 2.7 million tonnes/year.	2,762,239	11,403	3,881,898	12,842	5,399,358	13,884
	31,983,841	79,512	39,975,974	88,417	48,783,058	95,627
Reported as almost 500,000 tonnes/year of municipal waste generated, including HW.	566,202	5,663	884,585	6,767	1,252,664	7,888
8	3,989	11	9,038	13	11,933	15
4 (0.55 kg/person/day)	8,375,073	41,488	14,103,192	63,842	30,856,601	105,698
9 (11,562,600 tonnes/year collected in 2016 and a collection rate of 75.86 percent in 2012)	15,050,327	44,439	17,542,698	41,200	19,940,300	36,416
4 (1.6 kg/person/day)	5,413,453	9,270	6,802,059	11,055	8,571,552	13,164
	32,037,871	65,789	36,720,437	70,579	42,820,633	75,381
	263,726,732	322,180	311,039,297	354,712	359,887,136	389,592
1 (urban, 1.03 kg/person/day)	1,271,646	3,444	1,521,565	3,594	1,804,592	3,662
Reported as more than 4 million tonnes/year of MSW generated.	4,622,615	31,447	6,594,881	36,712	9,407,851	40,950
8	70,225	270	109,807	354	210,239	475
3; country reports 100 percent collection coverage.	10,093,925	31,568	11,693,608	36,750	15,756,898	41,585
4 (26,220 tonnes/day)	11,562,740	94,569	15,922,186	106,284	21,961,818	114,630
Calculated from waste generated in St. Thomas (65,000 tonnes/year) and in St. Croix (81,500 tonnes/year).	170,720	105	218,451	102	213,661	89

(Table continues on next page)

| Country or economy | Region | Income | Original year reported | | | Source |
			MSW generation	Population	Year	
West Bank and Gaza	MENA	LMIC	1,387,000	4,046,901	2012	GIZ and SWEEP-Net 2014d, 11
Yemen, Rep.	MENA	LMIC	4,836,820	27,584,213	2016	Al-Eryani 2017
Zambia	SSA	LMIC	2,608,268	14,264,756	2011	Edema, Sichamba, and Ntengwe 2012; Zambia, Central Statistical Office 2013
Zimbabwe	SSA	LIC	1,449,752	12,500,525	2002	GIZ 2013a, 2013b, 2013c, 2013d, 2014

Note: Year refers to year of data, unless otherwise specified.

Population for original year of data from the World Bank (2017b), except for Taiwan, China. Population for Taiwan, China, is from the Taiwan National Development Council (2015). Population for 2016 adjusted waste generation and 2030 and 2050 projected waste generation from UN DESA (2017).

For projection methodology, see box 2.1.

C&D = construction and demolition; EAP = East Asia and Pacific; ECA = Europe and Central Asia; HH = household; HIC = high-income country; HW = hazardous waste; ICI = institutional, commercial, and industrial; IW = industrial waste; kg = kilogram; LAC = Latin America and the Caribbean; LIC = low-income country; LMIC = lower-middle-income country; MENA = Middle East and North Africa; MSW = municipal solid waste; NA = North America; SAR = South Asia; SSA = Sub-Saharan Africa; UMIC = upper-middle-income country.

1. Calculated using an urban or city-specific daily or monthly MSW generation rate as a proxy for national urban generation; rural MSW generation is assumed to be 50 percent of urban or city rate; urban or city and value denoted in parentheses.
2. Personal communication.
3. Value represents amount collected.
4. Calculated using an average national MSW generation rate; value denoted in parentheses.
5. Value represents total solid waste generated, not only MSW.
6. One out of four countries in Sub-Saharan Africa for which no data were available at the country or city level. A regional estimate of waste generation per capita was calculated for the 44 other Sub-Saharan Africa countries; this regional per capita estimate was used to estimate total MSW generated for each of these four countries.
7. Value represents household waste only.
8. Calculated based on an urban regional per capita average of 1.3 kg/person/day and rural regional per capita average of 0.5 kg/person/day.
9. Calculated based on the amount of MSW collected and percentage of collection.

Comment	2016 adjusted		2030 projected		2050 projected	
	MSW generation	Population ('000s)	MSW generation	Population ('000s)	MSW generation	Population ('000s)
	1,628,920	4,791	2,768,338	6,739	5,618,921	9,704
1 (urban, 0.55–0.65 kg/person/day; rural 0.3–0.4 kg/person/day; averages used); 2	4,836,820	27,584	6,903,335	36,815	12,057,526	48,304
1 (Lusaka and Ndola, 0.72 kg/person/day)	3,114,269	16,591	5,239,016	24,859	11,185,099	41,001
1 (Chinhoyi, Gweru, Kariba, Kadoma, and Norton, 0.27 kg/person/day)	1,799,140	16,150	2,484,974	21,527	4,189,544	29,659

References

Abedini, Ali R. 2017. Solid waste management specialist and founder and CEO, ISWM Consulting Ltd. Personal communication with the World Bank.

ABRELPE (Brazilian Association of Public Cleaning and Special Waste Companies [Associação Brasileira de Empresas de Limpeza Pública e Resíduos Especiais]). 2015. "Overview of Solid Waste in Brazil 2015" ["Panorama dos Resíduos Sólidos no Brasil 2015"]. São Paulo.

Achankeng, Eric. 2003. "Globalization, Urbanization and Municipal Solid Waste Management in Africa." Paper presented at African Studies Association of Australasia and the Pacific, "African on a Global Stage," Adelaide, Australia.

ADB (Asian Development Bank). 2012. "Sector Assessment (Summary): Water and Other Municipal Infrastructure and Services." Solid Waste Management Improvement Project RRP UZB 45366, Asian Development Bank (ADB). Accessed April 18, 2017. https://www.adb.org/sites/default /files/linked-documents/45366-004-ssa.pdf.

———. 2013. *Solid Waste Management in Nepal: Current Status and Policy Recommendations.* Mandaluyong City, Philippines: Asian Development Bank.

———. 2014a. "Solid Waste Management in the Pacific: Papua New Guinea Country Snapshot." ADB Publication Stock No. ARM146612-2. Asian Development Bank, Manila. Accessed May 18, 2017. https://www .adb.org/sites/default/files/publication/42664/solid-waste-management -png.pdf.

———. 2014b. "Solid Waste Management in the Pacific: Solomon Islands Country Snapshot." ADB Publication Stock No. ARM146614-2. Asian Development Bank, Manila. Accessed May 9, 2017. https://www.adb .org/sites/default/files/publication/42662/solid-waste-management -solomon-islands.pdf.

———. 2014c. "Solid Waste Management in the Pacific: Tonga Country Snapshot." ADB Publication Stock No. ARM146616-2. Asian Development Bank, Manila.

Albania, INSTAT (Institute of Statistics). 2016. "Urban Solid Wastes in Albania." Institute of Statistics (INSTAT), Government of Albania, Tirana.

Al-Eryani, Muammer. 2017. Director of Solid Waste Management, Ministry of Local Administration (MoLA), Government of Republic of Yemen. Personal communication with the World Bank.

Al-Jayyousi, O. R. 2015. "National Action Plan." Ministry of Environment, Government of Jordan.

Amec Foster Wheeler. 2016. "National Solid Waste Management Strategy for the Cayman Islands: Final Report." Amec Foster Wheeler Environment and Infrastructure UK Limited, for the Government of Cayman Islands. http://ministryofhealth.gov.ky/sites/default/files/36082%20Strategy %20Final%20Report%2016229i1.pdf.

Angola, Ministry of Environment. 2012. "Strategic Plan for the Management of Urban Waste in Angola." ["Plano Estratégico para a Gestão de Resíduos Urbanos em Angola (PESGRUA)"]. Paper presented at the First Congress of Portuguese-Language Engineers [Primero Congresso dos Engenheiros de Língua Portuguesa], September 18.

Anon. 2015. "Integrated Solid Waste Management Plan, Bati Town, Amhara Regional State. Strategic Action Plan 2016–2025." November.

Armenia, National Statistical Service. 2017. "Waste Generation by Source, Hazardous Classes and Years." ArmState Bank Database. National Statistical Service, Government of Armenia. Accessed July 20, 2017. http://armstatbank.am.

Artelia Ville et Transport. 2014. "Solid Waste Characterization." R-11 Final Report. Solid Waste Management Project–Strategic and Technical Studies and Work Supervision. Addis Ababa City Government–Ethio–French Cooperation.

Azerbaijan, Ministry of Economy. 2017. "National Solid Waste Manegement Strategy Plan: Volume I (Main Report)." Aim Texas Trading, LLC–ICP Joint Venture, on behalf of Ministry of Economy, Government of Azerbaijan, Baku.

Barré, Juliette. 2014. "Waste Market in Urban Malawi: A Way Out of Poverty?" Master's thesis, Department of Urban and Rural Development, Faculty of Natural Resources and Agricultural Sciences, Swedish University of Agricultural Sciences, Uppsala, Sweden.

Barry, M. M. 2002. "Information on Household Waste Management in Conakry" ["Informations sur la gestion des ordures ménagères à Conakry"]. Cotonou Regional Workshop, "Shared Waste Management in African Cities" [Atelier régional de cotonou, "Gestion partagée des déchets dans les villes africaines"]. July 9–11.

Be'ah (Oman Environmental Service Holding Company S.A.O.C). 2016. "Transformation of Waste Management in Oman." Paper presented by M. al-Harthy at ISWA Energy Recovery WG Meeting, West Palm Beach, Florida, May 25.

Belarus, National Statistical Committee. 2017. National Statistical Committee, Government of Belarus. Personal communication with the World Bank.

BMDF (Bangladesh Municipal Development Fund). 2012. "Study on Municipal Solid Waste Management: Chittagong City Corporation, Rajshahi City Corporation, Rangpur Municipality, Patuakhali Municipality." BMDF, Dhaka.

Bolivia, DGGIRS. 2016. "National Infrastructure Program for Solid Waste Management, 2017–2020" ["Programa Nacional de Infraestructuras en Gestión Integral de Residuos Sólidos, 2017–2020"]. General Directorate of Integrated Solid Waste Management [Dirección General de Gestión Integral de Residuos Sólidos (DGGIRS)], Sub-Ministry of Potable Water and Basic Sanitation [Viceministerio de Agua Potable y Saneamiento Básico], Ministry of Environment and Water [Ministerio de Medio Ambiente y Agua], Government of Bolivia.

Bosnia and Herzegovina, BHAS (Agency for Statistics of Bosnia and Herzegovina). 2016. "First Release: Environment—Public Transportation and Disposal of Municipal Waste." Agency for Statistics of Bosnia and Herzegovina (BHAS), Bosnia and Herzegovina, Sarajevo. http://www .bhas.ba/saopstenja/2016/KOM_2015_001_01_BA.pdf.

Botswana, Statistics Botswana. 2011. "2011 Botswana Population and Housing Census." Statistics Botswana, Government of Botswana.

Brahim, R. 2017. "Household Waste Management in Tunisia" ["Gestion des déchets ménagers en Tunisie: Projets programmés et orientations des cahiers des charges de concession"]. National Waste Management Agency [Agence Nationale de Gestion des déchets (ANGed)], Ministry of Local Governments and of the Environment [Ministère des affaires locales et de l'environnement], Government of Tunisia.

Brunei, Department of Environment, Parks and Recreation. 2015. "Recycle 123 Handbook." Department of Environment, Parks and Recreation, Ministry of Development, Government of Brunei. http://www.env.gov .bn/Recyclers/Recycle%20123%20Handbook%204%20Nov%20 2015.pdf.

Burnside. 2014. "Preliminary Draft: Waste Characterization Report: Mangrove Pond Green Energy Complex." R. J. Burnside International Limited, for the Sanitation Service Authority, Government of Barbados. St. James, January.

Canada, Statistics Canada. 2016. "Waste Disposal by Source, Province and Territory." Statistics Canada, Government of Canada. Accessed November 10, 2017. http://www.statcan.gc.ca/tables-tableaux/sum-som/ l01/cst01/envir25a-eng.htm.

Caribbean Community Secretariat. 2013. "The CARICOM Environment in Figures 2009." Regional Statistics Programme, Caribbean Community Secretariat. http://www.caricomstats.org/Files/Publications/Environment 2009/CARICOMEnv2009.pdf.

Caribbean Development Bank. 2014. "Appraisal Report on Integrated Solid Waste Management Project—Grenada." Report No. AR 14/11 (BD 94/14). Paper presented at the Two Hundred and Sixty-Fourth Meeting of the Board of Directors, Caribbean Development Bank, Barbados, December 11.

CCAC (Climate and Clean Air Coalition). n.d. "Solid Waste Management City Profile: Lomé, Togo." Municipal Solid Waste Initiative. http://waste .ccac-knowledge.net.

Chen, K. 2016. "Taiwan: The World's Geniuses of Garbage Disposal." *Wall Street Journal*, May 17. https://www.wsj.com/articles/taiwan-the-worlds -geniuses-of-garbage-disposal-1463519134.

Chile, CONAMA (National Environmental Commission [Comisión Nacional de Medio Ambiente]). 2010. "First Report on Solid Waste Management in Chile" ["Primer Reporte del Manejo de Residuos Sólidos en Chile"]. CONAMA, Government of Chile.

Cissé, Sidi Mahamadou. 2015. Director of Sanitation, Government of Burkina Faso. Personal communication with the World Bank.

Costa Rica, Ministry of Health. 2016. "National Plan for Integrated Waste Management, 2016–2021." ["Plan Nacional para la Gestión Integral de Residuos, 2016–2021"]. Ministry of Health, Government of Costa Rica, San José.

Davis, J., S. Haase, and A. Warren. 2011. "Waste-to-Energy Evaluation: U.S. Virgin Islands." Technical Report NREL/TP-7A20-52308. National Renewable Energy Laboratory, Office of Energy Efficiency and Renewable Energy, U.S. Department of Energy, Washington, DC.

de Carvalho, J. M. C. 2013. "Elaboration of the Third International Conference on Sustainable Development in Small Island States in Development." National Report, United Nations Development Programme, Government of Cape Verde, Praia, July.

Delgerbayar, Badam. 2016. "Solid Waste Management in Mongolia." Paper presented at the Seventh Regional 3R Forum in Asia and the Pacific," Advancing 3R and Resource Efficiency for the 2030 Agenda for Sustainable Development," Adelaide, Australia, November 2–4. http:// www.uncrd.or.jp/content/documents/4136Country%20Presentation _Mongolia.pdf.

Dias, S., J. Vaz, and A. Carvalho. 2014. "Financing Waste Management in São Tomé and Principe." Paper presented at the Second International Africa Sustainable Waste Management Conference, Luanda, Angola, April 22–24.

Dominican Republic, Ministry of the Environment and Natural Resources. 2017. "Policy for Integrated Municipal Solid Waste Management" ["Política para la Gestión Integral de los Residuos Sólidos Municipales"]. Draft report. Ministry of the Environment and Natural Resources [Ministerio de Medio Ambiente y Recursos Naturales], Government of the Dominican Republic.

Ecuador, Ministry of Environment. 2018. National Program for Integrated Solid Waste Management [Programa Nacional para Gestión Integral de Residuos Sólidos], Ministry of Environment, Government of Ecuador. Personal communication with the World Bank. February.

Edema, Mojisola O., Victora Sichamba, and Felix W. Ntengwe. 2012. "Solid Waste Management: Case Study of Ndola, Zambia." *International Journal of Plant, Animal and Environmental Sciences* 2 (3): 248–55.

Eisted, Rasmus, and Thomas H. Christensen. 2011. "Waste Management in Greenland: Current Situation and Challenges." *Waste Management and Research* 29 (10): 1064–70.

Elbaroudi, N. O. M., S. E. M. Ahmed, and E. E. A. Adam. 2015. "Solid Wastes Management in Urban Areas: The Case of Khartoum State, Sudan." *Pinnacle Engineering and Technology* 3 (2): 616–24.

Energy Answers. 2012. "Materials Separation Plan." Energy Answers Arecibo LLC, Energy Answers International Inc.

Eurostat. 2017. "Municipal Waste by Waste Operations [env_wasmun]." Accessed April 25, 2017. http://ec.europa.eu/eurostat/web/waste /transboundary-waste-shipments/key-waste-streams/municipal-waste.

Ferrari, K., S. Cerise, R. Gamberini, B. Rimini, and F. Lolli. 2016. "An International Partnership for the Sustainable Development of Municipal Solid Waste Management in Guinea-Bissau, West Africa." Paper presented at the Twenty-First Summer School "Francesco Turco"—Industrial Systems Engineering. Naples, Italy, September 13–15.

Fiji, Department of Environment. 2011. "Fiji National Solid Waste Management Strategy, 2011–2014." Department of Environment, Ministry of Local Government, Urban Development, Housing and Environment, Government of Fiji. Accessed May 11, 2017. http://www .sprep.org/attachments/Fiji_NSWMS_2011-2014.pdf.

Francis, Shaka K. Y., Y. Higano, T. Mizunoya, and H. Yabar. 2015. "Preliminary Investigation of Appropriate Options for Leachate and Septage Treatment for the Caribbean Island of Antigua." Paper presented at the Fifty-Second Annual Meeting of the Japan Section of the Regional Science Association International (RSAI), Okayama University, Okayama, Japan, October 10–12. http://www.jsrsai.jp/Annual_Meeting /PROG_52/ResumeD/D02-4.pdf.

French Polynesia, DIREN (Directorate of the Environment). 2017. "Waste: Essential Data on Waste in French Polynesia" ["Les déchets: Les données essentielles sur les déchets en Polynésie française"]. Directorate of the Environment [La Direction de l'environnement (DIREN)], Government of French Polynesia, Papeete. http://www.environnement .pf/les-dechets.

Gibraltar, Ministry for the Environment. 2017. "Gibraltar Waste Management Plan 2013." Adopted December 2013. Revised December 2017. Environmental Agency Gibraltar and Department of the Environment, Ministry for the Environment, Government of Gibraltar. http://environmental-agency.gi/wp-content/uploads/2016/04/WasteManagementPlan2013.pdf.

GIZ (German Corporation for International Cooperation). 2013a. "Household and Commercial Solid Waste Characterisation Study: Kadoma City," by Hailey Johnson. GIZ, Bonn.

———. 2013b. "Household and Commercial Solid Waste Characterisation Study: Kariba," by Hailey Johnson. GIZ, Bonn.

————. 2013c. "Household, Commercial and Industrial Solid Waste Characterisation Study: Chinhoyi," by Hailey Johnson. GIZ, Bonn.

————. 2013d. "Household, Commercial and Industrial Solid Waste Characterisation Study: Norton," by Hailey Johnson. GIZ, Bonn.

————. 2014. "Household and Commercial Solid Waste Characterisation Study: Gweru," by Hailey Johnson. GIZ, Bonn.

————. 2015. "Preparation of an Assessment and Project Proposal for the Implementation of the Solid Waste Management Standards in Ethiopia." Urban Governance and Decentralisation Program. Implemented by RWA.

GIZ and SWEEP-Net. 2010a. "Country Report on the Solid Waste Management in Mauritania." German Corporation for International Cooperation [Deutsche Gesellschaft für Internationale Zusammenarbeit GmbH (GIZ)] and Regional Solid Waste Exchange of Information and Expertise Network in Mashreq and Maghreb Countries (SWEEP-Net), on behalf of the German Federal Ministry for Economic Cooperation and Development [Bundesministerium für wirtschaftliche Zusammenarbeit und Entwicklung (BMZ)].

————. 2010b. "Country Report on the Solid Waste Management in Syria." German Corporation for International Cooperation [Deutsche Gesellschaft für Internationale Zusammenarbeit GmbH (GIZ)] and Regional Solid Waste Exchange of Information and Expertise Network in Mashreq and Maghreb Countries (SWEEP-Net), on behalf of the German Federal Ministry for Economic Cooperation and Development [Bundesministerium für wirtschaftliche Zusammenarbeit und Entwicklung (BMZ)].

————. 2014a. "Country Report on the Solid Waste Management in Egypt." German Corporation for International Cooperation [Deutsche Gesellschaft für Internationale Zusammenarbeit GmbH (GIZ)] and Regional Solid Waste Exchange of Information and Expertise Network in Mashreq and Maghreb Countries (SWEEP-Net), on behalf of the German Federal Ministry for Economic Cooperation and Development [Bundesministerium für wirtschaftliche Zusammenarbeit und Entwicklung (BMZ)].

————. 2014b. "Country Report on the Solid Waste Management in Lebanon." German Corporation for International Cooperation [Deutsche Gesellschaft für Internationale Zusammenarbeit GmbH (GIZ)] and Regional Solid Waste Exchange of Information and Expertise Network in Mashreq and Maghreb Countries (SWEEP-Net), on behalf of the German Federal Ministry for Economic Cooperation and Development [Bundesministerium für wirtschaftliche Zusammenarbeit und Entwicklung (BMZ)]. April.

————. 2014c. "Country Report on the Solid Waste Management in Morocco." German Corporation for International Cooperation [Deutsche Gesellschaft für Internationale Zusammenarbeit GmbH (GIZ)] and Regional Solid Waste Exchange of Information and Expertise Network in Mashreq and Maghreb Countries (SWEEP-Net), on behalf of the German Federal Ministry for Economic Cooperation and Development [Bundesministerium für wirtschaftliche Zusammenarbeit und Entwicklung (BMZ)]. April.

———. 2014d. "Country Report on the Solid Waste Management in Occupied Palestinian Territories." German Corporation for International Cooperation [Deutsche Gesellschaft für Internationale Zusammenarbeit GmbH (GIZ)] and Regional Solid Waste Exchange of Information and Expertise Network in Mashreq and Maghreb Countries (SWEEP-Net), on behalf of the German Federal Ministry for Economic Cooperation and Development [Bundesministerium für wirtschaftliche Zusammenarbeit und Entwicklung (BMZ)].

Gogra, A. B., J. Yao, V. T. Simbay Kabba, E. H. Sandy, G. Zaray, S. P. Gbanie, and T. S. Bandagba. 2010. "A Situational Analysis of Waste Management in Freetown, Sierra Leone." *Journal of American Science* 6 (5): 124–35.

Greece, ELSTAT (Hellenic Statistical Authority). 2017. "Press Release: Waste Statistics, 2004–2014." ELSTAT, Government of Greece, Piraeus. January 31.

Guam. 2013. "Volume I: Guam Zero Waste Plan. Reaching for Zero: A Blueprint for Zero Waste in Guam." Government of Guam.

Guyana, Ministry of Communities. n.d. "Putting Waste in Its Place: A National Integrated Solid Waste Management Strategy for the Cooperative Republic of Guyana, 2017–2030—Part 1: Our Strategy." Ministry of Communities, Government of Guyana.

Honduras, DGA (General Directorate of Environmental Management). 2017. "Report on the Situation of Waste Management in Honduras 2016" ["Diagnóstico sobre la situación de la gestión de los residuos en Honduras 2016"]. General Directorate of Environmental Management [Dirección General de Gestión Ambiental (DGA)], Department of Energy, Natural Resources, Environment, and Mines [Secretaría de Energía, Recursos Naturales Ambiente y Minas (MiAmbiente)], Government of Honduras.

Hong Kong, Environmental Protection Department. 2016. "Hong Kong 2016 Waste Statistics: At a Glance." Waste Reduction Website, Environmental Protection Department (EPD), Government of Hong Kong SAR, China.

Iceland, Statistics Iceland. 2015. *Statistical Yearbook of Iceland 2015* [*Landshagir 2015*]. Statistics of Iceland III, 108 [Hagskýrslur Íslands III, 108]. Statistics Iceland [Hagstofa Íslands], Government of Iceland.

IDB (Inter-American Development Bank). 2010. "2020 Vision on Solid Waste: Strategic Solid Waste Management Plan, 2010–2020." IDB, Washington, DC.

———. 2012. "Strategic Plan for Solid Waste in Colombia" ["Plan Estratégico Sectorial de Residuos Sólidos de Colombia"], by G. A. Bernate. IDB, Washington, DC.

———. 2015. "Status of Solid Waste Management in Latin America and the Caribbean" ["Situación de la gestión de residuos sólidos en América Latina y el Caribe"]. IDB, Washington, DC.

IDB-AIDIS-PAHO (Inter-American Development Bank, Inter-American Association of Sanitary and Environmental Engineering, and Pan American Health Organization). 2011. "Report of the Regional Evalution of the Management of Municipal Solid Waste in Latin American and the Caribbean 2010" ["Informe de la Evaluación Regional del Manejo de Residuos Sólidos Urbanos en América Latina y el Caribe 2010"]. IDB, AIDIS, and PAHO.

Idrees, A. M., and K. P. McDonnell. 2016. "Feasibility Study of Applying Cynar Technology in the Gulf Cooperation Council Countries." *Biosystems and Food Engineering Research Review* 21: 88–91.

IMF (International Monetary Fund). 2004. "Djibouti: Poverty Reduction Strategy Paper." IMF Country Report No. 04/152, IMF, Washington, DC.

India, Ministry of Home Affairs. 2001. Census Data 2001. Office of the Registrar General and Census Commissioner, India, Ministry of Home Affairs, Government of India. http://www.censusindia.gov.in /2011-common/census_data_2001.html.

Indonesia, Ministry of Environment and Forestry and Ministry of Industry. 2016. "Indonesia Country Report on the Implementation of 3R Program." Paper presented at the Seventh Regional 3R Forum in Asia and the Pacific, "Advancing 3R and Resource Efficiency for the 2030 Agenda for Sustainable Development," Adelaide, Australia, November 2–4. http:// www.uncrd.or.jp/content/documents/456620-CB1-Indonesia.pdf.

Iraq, Ministry of Environment. 2015. "State of the Environment in Iraq 2015" [حالة البيئة في العراق لعام 2015]. Ministry of Environment, Government of Iraq.

Ireland, EPA (Environmental Protection Agency). 2014. "National Waste Report for 2012." EPA, Government of Ireland.

Isle of Man, Department of Infrastructure. n.d. "Waste Policy and Strategy, 2012 to 2022." Waste Management, Department of Infrastructure, Government of Isle of Man. https://www.gov.im/media/472034/waste _strategy.pdf.

Israel, Ministry of Environmental Protection. 2016. "New Waste Composition Survey." Ministry of Environmental Protection, Government of Israel. January 24. Accessed November 11, 2017. http://www.sviva .gov.il/English/env_topics/Solid_Waste/FactsAndFigures/Pages /WasteCompositionSurvey.aspx.

Isugi, Josephine, and Dongjie Niu. 2016. "Research on Landfill and Composting Guidelines in Kigali City, Rwanda, Based on China's Experience." *International Proceedings of Chemical, Biological and Environmental Engineering* 94: 62–68.

Jamaica, NSWMA (National Solid Waste Management Authority). n.d. "Waste Characterization and Per Capita Generation Rate Report 2013: The Metropolitan Wasteshed." Draft. Planning and Research Department, NSWMA, Government of Jamaica.

Japan, Ministry of the Environment. 2015. "Waste Treatment in Japan FY2015 Version" ["Nihonno haikibutsu syori heisei 27 nendo ban"]. Ministry of the Environment, Government of Japan. http://www.env.go .jp/recycle/waste_tech/ippan/h27/data/disposal.pdf.

Ji, L., S. Lu, J. Yang, C. Du, Z. Chen, A. Buekens, and J. Yan. 2016. "Municipal Solid Waste Incineration in China and the Issue of Acidification: A Review." *Waste Management and Research* 34 (4): 280–97.

Keohanam, Bounthong. 2017. Director, Urban Development Division, Department of Housing and Urban Planning, Ministry of Public Works and Transport, Government of Laos PDR. Personal communication with the World Bank. May 8.

Kho, P.-K., and S.-U. Lee. 2016. "Waste Resources Management and Utilization Policies of Korea." Korea Research Institute for Human Settlements, for the Ministry of Strategy and Finance, Government of South Korea. Accessed May 27, 2017. http://www.ksp.go.kr/publication /modul.jsp.

Korai, Muhammad Safar, Rasool Bux Mahar, and Muhammad Aslam Uqaili. 2017. "The Feasibility of Municipal Solid Waste for Energy Generation and Its Existing Management Practices in Pakistan." *Renewable and Sustainable Energy Reviews* 72: 338–53.

Kumar, Sunil, J. K. Bhattacharyya, A. N. Vaidya, Tapan Chakrabarti, Sukumar Devotta, and A. B. Akolkar. 2009. "Assessment of the Status of Municipal Solid Waste Management in Metro Cities, State Capitals, Class I Cities, and Class II Towns in India: An Insight." *Waste Management* 29 (2009) 883–95.

Kuwait, Partnerships Technical Bureau. 2014. "Building Kuwait's First Waste to Energy Plant: Kabd Municipal Solid Waste Project." Partnerships Technical Bureau, Government of Kuwait. http://www.recuwatt.com /pdf/aiduwaisan_manal.pdf.

Kyrgyzstan, NSC (National Statistical Committee of Kyrgyz Republic). 2016. "Annual Environmental Report, 2011–2015" [Статистический сборник 2011–2015]. NSC, Government of Kyrgyzstan. http://stat.kg/media /publicationarchive/194156d4-c806-4f02-9423-61ba3e85cce3.pdf.

Lesotho, Bureau of Statistics. 2006. "Total De Jure Population (Excluding Institutional Population) by District and Sex for 1996 and 2006." Bureau of Statistics, Government of Lesotho. http://www.bos.gov.ls/Census _Pre_Results_2006.htm.

———. 2013. "2012 Solid Waste, Water and Sanitation." Statistical Report No. 19. Bureau of Statistics, Government of Lesotho, Maseru.

Liechtenstein, Office of Statistics. 2018. "Liechtenstein in Figures 2018." Office of Statistics, Government of Liechtenstein. https://www.llv.li/files /as/liechtenstein-in-figures-2018.pdf.

Ludington, Gaïa. 2015. "Work Plan for Improved Consideration of Short-Lived Climate Pollutants in Household Solid Waste Management in Abidjan" ["Plan de travail pour une meilleure prise en compte des polluants climatiques à courte durée de vie dans la gestion des déchets solides ménagers à Abidjan"]. Gevalor, Orléans, France.

Macao SAR, China, DSEC (Statistics and Census Service). 2017. "Environmental Statistics 2016." Statistics and Census Service, Government of Macao SAR, China. http://www.dsec.gov.mo/Statistic /Social/EnvironmentStatistics/2016%E5%B9%B4%E7%92%B0%E5 %A2%83%E7%B5%B1%E8%A8%88.aspx?lang=en-US.

Macedonia, MAKStat (State Statistical Office of the Republic of Macedonia). 2017. "Municipal Waste, 2016." MAKStat, Government of Macedonia. http://www.stat.gov.mk/PrikaziSoopstenie_en.aspx?rbrtxt=80.

Maldives, Waste Management and Pollution Control Department and Ministry of Environment and Energy. 2017. Personal communication with the World Bank.

Mauritius, Ministry of Social Security, National Solidarity, and Environment and Sustainable Development. 2017. "SWMD-Solid Waste Management in Mauritius: The Solid Waste Management Division." Environment and Sustainable Development Division, Ministry of Social Security, National Solidarity, and Environment and Sustainable Development, Government of Mauritius. Accessed December 5, 2017. http://environment.govmu .org/English/Pages/swmd/SWMD-Solid-Waste-In-Mauritius.aspx.

Mbue, N. I., D. Bitondo, and R. Azibo Balgah. 2015. "Municipal Solid Waste Generation, Composition, and Management in the Douala Municipality, Cameroon." *Journal of Environment and Waste Management* 2 (4): 91–101.

Mexico, SEMARNAT (Secretariat of Environment and Natural Resources). 2016. "Report on the Situation of the Environment in Mexico" ["Informe de la Situación del Medio Ambiente en México"]. General Directorate of Environmental Statistics and Information [Dirección General de Estadística e Información Ambiental], SEMARNAT, Government of Mexico.

Modak, Prasad, Agamuthu Pariatamby, Jeffrey Seadon, Perinaz Bhada-Tata, Guilberto Borongan, Nang Sian Thawn, and Ma Bernadeth Lim. 2017. *Asia Waste Management Outlook*, edited by P. Modak. Nairobi: United Nations Environment Programme, Asian Institute of Technology, and International Solid Waste Association.

Mohee, R., S. Mauthoor, Z. M. A. Bundhoo, G. Somaroo, N. Soobhany, and S. Gunasee. 2015. "Current Status of Solid Waste Management in Small

Island Developing States: A Review" [Supplementary Material]. *Waste Management* 43: 539–49.

Moldova, Statistica Moldovei (National Bureau of Statistics of the Republic of Moldova). 2016. Natural Resources and the Environment in the Republic of Moldova: Collected Statistics [Resursele naturale şi mediul în Republica Moldova: Culegere statistică / Природные ресурсы и окружающая среда в Республике Молдова: Статистический сборник]. Statistica Moldovei, Government of Moldova, Chişinău.

Mombo, Jean-Bernard, and Mesmin Edou. 2005. "Urban Solid Waste Management in Gabon" ["La gestion des déchets solides urbains au Gabon"]. *Geo-Eco-Trop* 29: 89–100.

Naquin, P. 2016. "Design and Implementation of a Quantification and Characterization Campaign for Household Waste in the Territory of AITOM le Marien (Cap Haïtien–Limonade–Quartier Morin)" ["Conception et réalisation d'une campagne de quantification et de caractérisation de déchets ménagers sur le territoire de l'AITOM le Marien (Cap Haïtien–Limonade–Quartier Morin)"]. Contract ATN/OC–5 400 – HA. Inter-American Development Bank, Washington, DC.

Nepal, SWMTSC (Solid Waste Management Technical Support Center). 2017. "Solid Waste Management Baseline Study of 60 New Municipalities: Finale Report." By the Engineering Study and Research Centre (P) Ltd. (ESRC). SWMTSC, Ministry of Federal Affairs and Local Development, Government of Nepal.

Nguyen, H. H., S. Heaven, and C. Banks. 2014. "Energy Potential from the Anaerobic Digestion of Food Waste in Municipal Solid Waste Stream of Urban Areas in Vietnam." *International Journal of Energy and Environmental Engineering* 5: 365–74.

Nizami, A.-S. 2015. "Recycling and Waste-to-Energy Prospects in Saudi Arabia." BioEnergy Consult, November 10. Accessed April 28, 2017. http://www.bioenergyconsult.com/recycling-waste-to-energy-saudi-arabia.

Nordic Competition Authorities. 2016. "Competition in the Waste Management Sector: Preparing for a Circular Economy." Nordic Competition Authorities, Konkurrensverket (Sweden), Konkurransetilsynet (Norway), Kilpailu ja Kuluttajavirasto (Finland), Samkeppniseftirlitid (Iceland), Konkurrence og Forbrugerstyrelsen (Denmark), Kappingareftirlitid (Faroe Islands), and Forbruger og Konkurrencestyrelsen (Greenland). http://www.konkurrens-verket.se/globalassets/publikationer/nordiska/nordic-report-2016_waste-management-sector.pdf.

OECD (Organisation for Economic Co-operation and Development). 2018. "Municipal Waste, Generation and Treatment." OECD.Stat. OECD, Paris. https://stats.oecd.org/Index.aspx?DataSetCode=MUNW.

Oguntoyinbo, O. O. 2012. "Informal Waste Management System in Nigeria and Barriers to an Inclusive Modern Waste Management System: A Review." *Public Health* 126 (5): 441–47.

Okot-Okumu, James. 2012. "Solid Waste Management in African Cities—East Africa." In *Waste Management: An Integrated Vision*, edited by Luis Fernando Marmolejo Rebellon, 3–20. London: InTechOpen.

Okot-Okumu, James, and Richard Nyenje. 2011. "Municipal Solid Waste Management under Decentralisation in Uganda." *Habitat International* 35 (2011): 537–43.

Omran, A., M. Altawati, and G. Davis. 2017. "Identifying Municipal Solid Waste Management Opportunities in Al-Bayda City, Libya." *Environment, Development and Sustainability* 20 (4): 1597–613.

Ouamane, Karim. 2017. General director, Waste Management Agency, Government of Algeria. Personal communication with the World Bank.

Particip. 2015. "National Waste Management Strategy, 2016–2030 (Draft): Development of Legislation for Waste Management as Part of the EU-Georgia Association Agreement." Prepared by Particip, with the financial assistance of the European Commission, Freiburg, Germany.

Pattle Delamore Partners Ltd. 2015. "Assessment of Status and Options for Solid Waste Management on Majuro Atoll." Prepared by Tristan Bellingham. SPREP (Secretariat of the Pacific Regional Environment Programme), Apia, Samoa. Accessed May 10, 2017. https://www.sprep. org/attachments/pacwaste/A02753600R001Final_Rev2_with_ Appendices.pdf.

Peru, Ministry of Environment. 2014. "Sixth National Report on Municipal and Non-Municipal Management of Solid Waste" ["Sexto informe nacional de residuos sólidos de la gestión del ámbito municipal y no municipal 2013"]. Ministry of Environment, Government of Peru, Lima.

———. 2016. "National Plan for Integrated Solid Waste Management, 2016–2024" ["Plan Nacional de Gestión Integral de Residuos Sólidos, 2016–2024"]. Ministry of Environment, Government of Peru, Lima.

Philippines, NSWMC (National Solid Waste Management Commission). 2017. "Solid Waste Management Dashboard." NSWMC, Government of Philippines. Acessed April 20, 2017. http://119.92.161.4/nswmc4 /default3.aspx.

Phuntsho, S., I. Dulal, D. Yangden, S. Herat, H. Shon, and S. Vigneswaran. U. M. Tenzin. 2010. "Studying Municipal Solid Waste Generation and Composition in the Urban Areas of Bhutan." *Waste Management and Research* 28 (6): 545–51.

PricewaterhouseCoopers Aruba. 2014. "Environmental Sustainability Ranking: Determining and Fortifying Aruba's Position in the Caribbean." Version 1.1. PricewaterhouseCoopers Aruba, October 28. https://www .pwc.com/an/en/publications/assets/pwc-environmental-sustainability -ranking-positioning-and-fortifying-aruba-in-caribbean.pdf.

Puopiel, F. 2010. "Solid Waste Management in Ghana: The Case of Tamale Metropolitan Area." MSc thesis, Kwame Nkrumah University of Science and Technology, Kumasi, Ghana.

Qatar, MDPS (Ministry of Development Planning and Statistics). 2014. "Environment Statistics Annual Report 2013." MDPS, Government of Qatar. http://www.mdps.gov.qa.

Rebelde, Juventud. 2007. "The 100th Street Landfill" ["El Vertedero de la Calle 100"]. July 8.

San Marino, AASS (Autonomous State Company for Public Services). 2016. "Collection Data" ["Dati di raccolta"]. AASS, Government of San Marino. http://www.aass.sm/site/home/ambiente/dati-di-raccolta/2016.html.

Sanneh, E. S., Allen H. Hu, Y. M. Chang, and Edrisa Sanyang. 2011. "Introduction of a Recycling System for Sustainable Municipal Solid Waste Management: A Case Study on the Greater Banjul Area of the Gambia." *Environment, Development and Sustainability* 13: 1065–80.

Seanama Conservation. 2012. "Botswana National Report for the United Nations Conference on Sustainable Development (Rio+20)." Seanama Conservation Consultancy (Pvt.), Ltd.

Senegal, UCG (Coordinating Unit for Solid Waste Management). 2016. "National Report: Report of the National Campaign for the Characterization of Household and Integrated Waste (2016)" ["Rapport national: Rapport de la campagne nationale de caractérisation des ordures ménagères et assimilées (2016)"]. UCG, Ministry of Local Governance, Development, and Planning, Government of Senegal.

SIDS DOCK (Small Island Developing States). 2015. "Toward the Development of a Caribbean Regional Organic Waste Management Sub-Sector: Development of a Caribbean Regional Organic Waste Management Conversion Sub-Sector to Increase Costal Resilience and Climate Change Impacts and Protect Fresh Water Resources." Draft 1.0. SIDS DOCK Secretariat and Caribbean Community Climate Change Centre.

Simos, J., and E. de Leeuw. 2017. "Healthy Cities in Africa: A Continent of Difference." In *Healthy Cities: The Theory, Policy, and Practice of Value-Based Urban Planning*, edited by E. de Leeuw and J. Simos, 89–132. New York: Springer.

Singapore, NEA (National Environment Agency). 2017. "Waste Statistics and Overall Recycling." NEA, Government of Singapore. http://www.nea.gov.sg/energy-waste/waste-management/waste-statistics-and-overall-recycling.

South Africa, Department of Environmental Affairs. 2012. "National Waste Information Baseline Report: Draft." Department of Environmental Affairs, Government of South Africa, Pretoria, November 14.

SPREP (Secretariat of the Pacific Regional Environment Programme). 2016. "Cleaner Pacific 2025: Pacific Regional Waste and Pollution Management Strategy, 2016–2025." SPREP, Apia, Samoa. https://sustainabledevelopment.un.org/content/documents/commitments/1326_7636_commitment_cleaner-pacific-strategy-2025.pdf.

Sri Lanka, Ministry of Mahaweli Development and Environment. 2016. "Comprehensive Integrated Solid Waste Management Plan for Target Provinces in Sri Lanka." Ministry of Mahaweli Development and Environment, Government of Sri Lanka.

States of Guernsey. 2017. "Guernsey Facts and Figures 2017." States of Guernsey Data and Analysis. https://www.gov.gg/CHttpHandler.ashx?id =110282andp=0.

States of Jersey. 2018. "Waste Management s\Statistics." Jersey in figures. Environmental statistics. https://www.gov.je/Government/JerseyInFigures /Environment/pages/wastemanagement.aspx.

Sterviinou, L. n.d. "What to Do with Our Waste in Saint Martin?" ["Quoi faire de nos déchets à Saint-Martin?"]. MAXImini.com.

St. Lucia, SLSWMA (St. Lucia Solid Waste Management Authority). 2015. "Saint Lucia Solid Waste Management Authority Annual Report, April 2014–March 2015." Saint Lucia Solid Waste Management Authority, Government of Saint Lucia.

SWANA Haiti Response Team. 2010. "Municipal Solid Waste Collection Needs in Port-au-Prince, Haiti: Position Paper." Solid Waste Association of North America (SWANA).

Taiwan National Development Council. 2015. "Taiwan Statistical Data Book." National Development Council, Government of Taiwan, China. July. https://www.ndc.gov.tw/en/News_Content.aspx?n=607ED343456 41980andsms=B8A915763E3684ACands=A016E84591034DC5.

Tajikistan, Tajstat. 2017. "Time Series Data on Indicators for the Period 1990–2013: Waste Generation" [Временные ряды данных по показателям за период 1990-2013 гг., Образование отходов]. Tajstat, Government of Tajikistan. Accessed November 10, 2017. http://www.gksintranet.tj /ecostat/Otkhod.html.

Talma, Elme, and Michele Martin. 2013. "The Status of Waste Management in Seychelles." Sustainability for Seychelles. GEF (Global Environment Facility), SGP (The GEFs Small Grants Programme), and United Nations Development Programme.

Tanzania, NBS (National Bureau of Statistics) and OCGS (Office of Chief Government Statistician). 2014. "Basic Demographic and Socio-Economic Profile." NBS, Ministry of Finance, Dar es Salaam, and Office of Chief Government Statistician, Ministry of State, President's Office, State House and Good Governance, Zanzibar, Government of Tanzania.

Tas, Adriaan, and Antoine Belon. 2014. "A Comprehensive Review of the Municipal Solid Waste Sector in Mozambique: Background Documentation for the Formulation of Nationally Appropriate Mitigation Actions in the Waste Sector in Mozambique." Carbon Africa Limited and Mozambican Recycling Association [Associação Moçambicana de Reciclagem]. Maputo and Nairobi.

Tekes (Finnish Funding Agency for Technology and Innovation). 2013. "Future Watch Report: Future of Waste Management in Russian Megacities." Tekes—The Finnish Funding Agency for Technology and Innovation. December. Accessed April 25, 2017. https://www.tekes.fi /globalassets/julkaisut/future_of_waste_management_in_russian _megacities.pdf.

Thailand, PCD (Pollution Control Department). 2015. "Thailand State of Pollution Report 2015." PCD No. 06-062. PCD, Ministry of Natural Resources and Environment, Government of Thailand. Accessed April 1, 2016. http://infofile.pcd.go.th/mgt/PollutionReport2015_en.pdf.

Thein, M. 2010. "GHG Emissions from Waste Sector of INC of Myanmar." Paper presented at the Eighth Workshop on GHG Inventories in Asia (WGIA8), Vientiane, Lao PDR, July 13–16.

Timor-Leste, Ministry of Commerce, Industry and the Environment. 2016. "Country Report (Draft): Timor-Leste." Paper presented at the Seventh Regional 3R Forum in Asia and the Pacific, Adelaide, Australia, November 2–4. Accessed May 25, 2017. http://www.uncrd.or.jp/content /documents/4056Country%20Report_Timor%20Leste.pdf.

Treasure, A. S.-O. n.d. "Comparison of Municipal Solid Waste Characterization and Generation Rates in Selected Caribbean Territories and Their Implications."

Trinidad and Tobago, EMA (Environmental Management Authority). n.d. "State of the Environment Report, 2011." EMA, Government of Trinidad and Tobago.

Tshitala Kalula, P. 2016. "A Study of the Solid Waste Management Sector in the DRC: The Case of the City of Kinshasa, from 25 August to 19 December 2016" ["État des lieux de la gestion des déchets solides en République Démocratique du Congo: Cas de la ville de Kinshasa, situation du 25 août au 19 décembre 2016"].

Ukraine, SSC (State Statistics Service of Ukraine). 2017. "Management of Household and Similar Wastes in Ukraine for 2011–2015." SSC, Government of Ukraine. Accessed June 1, 2017. https://ukrstat.org/en/ operativ/operativ2013/ns_rik/ns_e/pzppv_2013_e.htm.

UNCRD (United Nations Centre for Regional Development) and IGES (Institute for Global Environmental Strategies). 2017. "State of 3Rs in Asia and the Pacific, Country Report: Malaysia." Secretariat of the Regional 3R Forum in Asia and the Pacific, UNCRD and IGES, Kamiyamaguchi, Japan.

UN DESA (United Nations, Department of Economic and Social Affairs). 2014a. "Population of Urban Agglomerations with 300,000 Inhabitants or More in 2014, by Country, 1950–2030 (Thousands)" (UN Doc. No. POP/DB/WUP/Rev.2014/1/F12). In *World Urbanization Prospects: The 2014 Revision*. New York: Population Division, UN DESA.

———. 2014b. "World Urbanization Prospects: The 2014 Revision." UN Doc. No. ST/ESA/SER.A/366. Population Division, UN DESA, New York.

———. 2017. "World Population Prospects: The 2017 Revision." Population Division, UN DESA, New York.

UNDP (United Nations Development Programme). 2014. "Lebanon Environmental Assessment of the Syrian Conflict and Priority Interventions." UNDP, New York. http://www.undp.org/content/dam/lebanon/docs/Energy%20and%20Environment/Publications/EASC-WEB.pdf.

UNECA-UNEP-UNIDO-ARSCP (United Nations Economic Commission for Africa, United Nations Environment Programme, United Nations Industrial Development Programme, Africa Roundtable on Sustainable Consumption and Production. 2011. "Sustainable Consumption and Production for Sustainable Growth and Poverty Reduction" (SDSRA Volume III). https://www.uneca.org/sites/default/files/PublicationFiles/sdra3.pdf.

UNEP (United Nations Environment Programme). 2007. "Assessment of Solid Waste Management in Liberia." Post-Conflict and Disaster Management Branch, UNEP, in collaboration with the Environmental Protection Agency, Government of Liberia.

———. 2013. "Municipal Solid Waste Composition Analysis Study: Juba, South Sudan." UNEP, Nairobi.

UNICEF-India, Ministry of Rural Development. 2008. "Solid and Liquid Waste Management in Rural Areas—A Technical Note." Ministry of Rural Development, Department of Drinking Water Supply.

UN OCHA (Office for the Coordination of Humanitarian Affairs). 2014. "Central African Republic: WASH Update, Mai 2014." UN OCHA, New York.

UNSD (United Nations Statistics Division). 2016. "Environmental Indicators: Waste: Municipal Waste Collected." UNSD, New York, November. http://unstats.un.org/unsd/ENVIRONMENT/qindicators.htm.

U.S. EPA (Environmental Protection Agency). 2014. "Advancing Sustainable Materials Management: 2014 Fact Sheet: Assessing Trends in Material Generation, Recycling, Composting, Combustion with Energy Recovery and Landfilling in the United States." Office of Land and Emergency Management, EPA, Washington, DC.

Venezuela, INE (National Statistical Institute). 2013. "Generation and Management of Solid Waste in Venezuela, 2011–2012" ["Generación y manejo de residuos y desechos sólidos en Venezuela, 2011–2012"]. Report No. 3. INE, Government of Venezuela.

WaterAid. 2015a. "Solid Waste Generation Rate and Characterization Study for Bule Hora Town." Getaneh Gebre Consultancy Service, Addis Ababa.

———. 2015b. "Solid Waste Generation Rate and Characterization Study for Gerbe Guracha Town." Getaneh Gebre Consultancy Service, Addis Ababa.

World Bank. 2014. "Results-Based Financing for Municipal Solid Waste." Urban Development Series Knowledge Papers No. 20. Global Urban and DRM Unit, World Bank, Washington, DC.

———. 2015a. "Addressing Challenges of Municipal Solid Waste Management in Moroni (P145255)." Output-Based Aid Concept Note Executive Summary, World Bank, Washington, DC.

———. 2015b. "Assessment of the Integrated Management of Municipal Solid Waste in Argentina: Collection, Generation and Analysis of Data: Collection, Sweeping, Transfer, Treatment and Final Disposal of Municipal Solid Waste." [Diagnóstico de la Gestión Integral de Residuos Sólidos Urbanos en la Argentina: Recopilación, generación y análisis de datos: Recolección, barrido, transferencia, tratamiento y disposición final de Residuos Sólidos Urbanos]. World Bank, Washington, DC.

———. 2016a. "Integrated Urban Development and Resilience Project for Greater Antananarivo (P159756): Project Information Document/ Integrated Safeguards Data Sheet (PID/ISDS)." Concept Stage. Report No. PIDISDSC17608. World Bank, Washington, DC.

———. 2016b. "Rapid Assessment of Kabul Municipality's Solid Waste Management System." Report No. ACS19236. Environment and Natural Resources Global Practice, South Asia Region, World Bank. June.

———. 2017a. "Financial Analysis in Support of the Dominica Solid Waste Management Corporation Modernization Project." Solid Waste Management and Disaster Risk Management (GFDRR), Social, Urban, Rural and Resilience Global Practice, World Bank, Washington, DC.

———. 2017b. Population, total (database). World Bank, Washington, DC. https://data.worldbank.org/indicator/SP.POP.TOTL?locations=LK.

———. n.d. "Legal, Institutional, Financial Arrangement and Practices of Solid Waste Management Sector in Kazakhstan." World Bank, Washington, DC.

Zambia, Central Statistical Office. 2013. "Population and Demographic Projections, 2011–2035." Central Statistical Office, Government of Zambia. http://www.zamstats.gov.zm.

Zoï Environment Network. 2013. "Waste and Chemicals in Central Asia: A Visual Synthesis." Zoï Environment Network, with support from the Swiss Federal Office for the Environment. Accessed April 27, 2017. http://wedocs.unep.org/handle/20.500.11822/7538.

Zuilen, L. F. 2006. "Planning of an Integrated Solid Waste Management System in Suriname: A Case Study in Greater Paramaribo with Focus on Households." PhD thesis, Department of Soil Management, Faculty of Bioscience Engineering, Ghent University, Ghent, Belgium.

Waste Treatment and Disposal by Country or Economy

Country or economy	Region	Income	Open dump	Landfill unspecified	Controlled landfill	Sanitary landfill	Recycling	Com- posting	Anaerobic digestion	Incin- eration
Algeria	MENA	UMIC			2.0	89.0	8.0	1.0		
Andorra	ECA	HIC								52.1
Antigua and Barbuda	LAC	HIC			98.7			0.1		
Argentina	LAC	UMIC	22.6		8.9	62.5	6.0			
Armenia	ECA	LMIC	100.0							
Aruba	LAC	HIC					11.0			
Australia	EAP	HIC		48.9			42.1			9.8
Austria	ECA	HIC		3.0			25.7	31.2		37.9
Azerbaijan	ECA	UMIC	100.0							
Bahrain	MENA	HIC		92.0			8.0			
Bangladesh	SAR	LMIC						5.3		

Advanced thermal treatment	Water-ways	Other	Un-accounted for	Year(s)	Source	Comment
				2016, 2013	**CLF, SLF:** Ismail 2017 **RE, CM:** GIZ and SWEEP-Net 2014a	1
			47.9	2012	UNSD 2016	**UA:** Does not include landfilling, recycling, or composting, as values for those are 0.
0.1			1.1	2014, 2011	**CLF:** Gore-Francis 2013 **CM, UA, WW:** Antigua and Barbuda, Statistics Division 2014, 36	**CLF:** Cooks Sanitary Landfill and Civic Amenities Site is referred to as a "sanitary landfill"; however, based on the performance audit report by the National Solid Waste Management Authority, 2013, it is run as a CLF; CLF estimated by subtracting uncollected waste (composted or thrown in waterways) from 100 percent of the waste. **CM:** Calculated by the population composting compared with total population disposing of garbage by various means from questionnaire on Population by Parish by Waste Disposal Method. **WW:** Calculated by the population dumping waste in river, sea, or pond compared with total population disposing of garbage by various means from questionnaire on Population by Parish by Waste Disposal Method. **UA:** Calculated by the population burning (0.23 percent), burying (0.04 percent), not stated (0.56 percent), and other (0.31 percent) compared with total population disposing of garbage by various means from questionnaire on Population by Parish by Waste Disposal Method.
				2010	Argentina SIDSA 2015, 80	
				2012	UNSD 2016	New sanitary landfills are being constructed as of 2018.
			89.0	2013	Pricewaterhouse-Coopers Aruba 2014	
				2015	Australia, Department of the Environment and Energy 2017, vii	**LF:** Calculated based on amount of MSW disposed of compared with amount generated. **RE:** Calculated based on amount of MSW recycled compared with amount generated. **IN:** Calculated based on amount of MSW recovered for energy compared with amount generated.
			2.2	2015	Eurostat 2017	**CM:** 2
				2015	Azerbaijan, Ministry of Economy 2017, 105	Based on the statement "All of Azerbaijan's disposal sites, other than the new systems within the Baku metropolitan area, is by open dumping."
				2012	Al Sabbagh et al. 2012	**RE:** Value for recycling and material recovery rate.
			94.8	2011	**OD:** Enayetullah, Sinha, and Khan 2005 **CM:** ADB 2011	**OD:** Most common method of waste disposal in Bangladesh.

(Table continues on next page)

Country or economy	Region	Income	Open dump	Landfill unspecified	Controlled landfill	Sanitary landfill	Recycling	Com- posting	Anaerobic digestion	Incin- eration
Barbados	LAC	HIC			90.0		9.0			
Belarus	ECA	UMIC	7.1		76.9		16.0			
Belgium	ECA	HIC		0.9			34.3	19.1		43.4
Belize	LAC	UMIC	66.0			34.0				
Benin	SSA	LIC					25.0			
Bermuda	NA	HIC			12.2		2.0	18.3		67.6
Bhutan	SAR	LMIC			98.0		0.9	1.4		
Bolivia	LAC	LMIC	55.5		0.0	31.9	12.1	0.4		
Bosnia and Herzegovina	ECA	UMIC	41.8		8.6	24.1				
Botswana	SSA	UMIC					1.0			
Brazil	LAC	UMIC	15.6		21.9	53.3	1.4	0.2		

Advanced thermal treatment	Water-ways	Other	Un-accounted for	Year(s)	Source	Comment
			1.0	2012, 2015	Riquelme, Méndez, and Smith 2016	**CLF:** Estimated based on total generation minus the amount uncollected; the main landfill in use is the Mangrove facility, which is considered a CLF. **RE:** Represents only a portion of waste that is recycled by the Sustainable Barbados Recycling Centre; the actual amount of HH waste that is finally recycled is not known. **CM:** 3. **UA:** Calculated based on difference between total waste generation and sum of waste landfilled and recycled.
				2016	**OD, CLF:** Belarus, Ministry of Housing and Utility 2017b **RE:** Belarus, Ministry of Housing and Utility 2017a	1
			2.3	2015	Eurostat 2017	**CM:** 2
				2012	**OD, SLF:** IDB 2015 **CM:** IDB 2013, 18	**CM:** 3 **RE:** 4 **OD:** Reported as "inadequate disposal of waste" in source, which includes open dumps, open burning, and other forms of final disposal (bodies of water, animal feed, and so on), of which most is assumed to be openly dumped.
			75.0	2005	AFED 2008, 18	
				2012	UNSD 2016	**CLF:** 5 **IN:** Value represents IN and ATT.
				2016	Bhutan, National Environment Commission 2016	**CLF:** Majority of waste is dumped at a Memelakha controlled landfill in Thimphu; it has soil cover and compaction of waste on a regular basis.
				2015	Bolivia, MMAyA/ VAPSB/DGGIRS 2016	**OD:** Includes remainder of waste not processed through formal collection or informal recycling. **CLF, SLF:** SLF includes those constructed as such and in good operation; CLF are those constructed as SLF but not operating well; abandoned landfills excluded. **RE:** Includes informal recycling based on estimates and formal recycling based on inventory of existing plants. **CM:** Based on inventory of existing plants.
		0.0	25.6	2015	Bosnia and Herzegovina, BHAS 2016, 1	**CLF:** Reported as controlled landfill because there is no landfill gas management. **Other:** 6
			99.0	2005	AFED 2008, 18	
			7.6	2015, 2014	**OD, CLF, SLF:** ABRELPE 2015, 23 **IN:** ABRELPE 2015, 69 **RE:** UFPE 2014, 84 **CM:** Brazil SNIS 2017, 145	**CM:** Data refer to percentage of waste sent to open dumps, sanitary landfills, compost, and sorting plants in participating municipalities.

(Table continues on next page)

Country or economy	Region	Income	Open dump	Landfill unspecified	Controlled landfill	Sanitary landfill	Recycling	Com- posting	Anaerobic digestion	Incin- eration
British Virgin Islands	LAC	HIC								80.3
Brunei Darussalam	EAP	HIC				70.0		2.0		
Bulgaria	ECA	UMIC		66.2			19.0	10.3		2.8
Burkina Faso	SSA	LIC	59.0	17.0			12.0			
Cambodia	EAP	LMIC								
Cameroon	SSA	LMIC	80.3			19.3	0.4			
Canada	NA	HIC		72.3			20.6	4.1		3.0
Cayman Islands	LAC	HIC					21.0			
Channel Islands	ECA	HIC			39.2		28.4	15.9		16.4
Chile	LAC	HIC	8.4			85.3	0.4	0.4		0.1
China	EAP	UMIC	8.2	60.2				3.0		29.8
Colombia	LAC	UMIC	4.0			89.0	17.2			
Congo, Dem. Rep.	SSA	LIC					4.9			

Advanced thermal treatment	Water-ways	Other	Un-accounted for	Year(s)	Source	Comment
			19.7	2005	UNSD 2016	
			28.0	2014	Shams, Juani, and Guo 2014	
			1.7	2015	Eurostat 2017	
			12.0	2009, 2005	**OD, LF, UA:** IMF 2012 **CM, AD:** Cissé 2015 **RE:** AFED 2008, 18	**CM:** 3 **AD:** 7 (There is a small-scale biogas facility in Ouagadougou). **UA:** Includes some open burning.
	17.5		82.5	2004	Patriamby and Tanaka 2014, 82	8 **OD:** 9 **SLF:** There is a sanitary landfill in Phnom Penh. **Other:** Includes open burning (15 percent) and other unspecified methods in urban areas (2.5 percent); other methods (unspecified) in suburban areas is 15 percent (not included in figure).
				2012	**OD, SLF:** UNFCCC 2014 **CM, Other:** Armel 2017 **RE:** UNSD 2016	**CM:** 3 **OD:** Proportion of waste that is not landfilled or recycled is dumped. **Other:** 10
				2008, 2007	**LF, RE, IN:** Canada, Statistics Canada 2012 **CM:** van der Werf and Cant 2007	**LF:** In total, 25,871,310 tonnes are disposed of by landfill and incineration, amounting to 75.3 percent of waste; 3 percent is incinerated, so total landfill amount is 72.3 percent. **RE:** In total, 8,473,257 tonnes were diverted to recycling and composting. According to a 2007 article, 17 percent of organic waste is composted, which in total is about 4.1 percent of waste; thus total recycled is approximately 20.6 percent. **CM:** 17 percent of organic waste is composted, which in total is about 4.1 percent of waste. **IN:** Seven municipal incineration plants in Canada.
			79.0	2013	**LF:** Amec Foster Wheeler 2016 **RE:** Pricewaterhouse-Coopers Aruba 2014	**LF:** 11 (3 landfills).
				2016	States of Guernsey 2017	
		0.0	5.3	2009	Chile, CONAMA 2010, 59	14 **IN:** Includes both with and without energy recovery.
				2014, 2011	**OD, LF, IN:** Modak et al. 2017, 215 **CM:** Takeda, Wang, and Takaoka 2014, 35	**CM:** Includes biological treatment and other treatment technologies. Informal recycling is estimated to be 15.8% nationally based on a World Bank 2011 study.
				2011	**OD, SLF:** IDB 2012, 28 **RE:** IDB 2015, 3	
			95.1	2005	**RE:** AFED 2008, 18 **Other:** Kalula 2016	**SLF:** 12

(Table continues on next page)

Country or economy	Region	Income	Open dump	Landfill unspecified	Controlled landfill	Sanitary landfill	Recycling	Com-posting	Anaerobic digestion	Incin-eration
Congo, Rep.	SSA	LMIC					26.2			
Costa Rica	LAC	UMIC	9.1		23.5	67.5	1.3			
Côte d'Ivoire	SSA	LMIC					3.0			
Croatia	ECA	UMIC		79.8			16.3	1.7		
Cuba	LAC	UMIC	42.2		30.7		9.5			
Curaçao	LAC	HIC					2.0			
Cyprus	ECA	HIC		74.5			13.3	4.6		
Czech Republic	ECA	HIC		52.6			25.5	4.2		17.7
Denmark	ECA	HIC		1.1			27.3	19.0		52.6
Dominica	LAC	UMIC			94.0					
Dominican Republic	LAC	UMIC	72.6			0.1	8.2			

Advanced thermal treatment	Water-ways	Other	Un-accounted for	Year(s)	Source	Comment
			73.8	2005	**OD, CM:** Guillaume, Château, and Tsitsikalis 2015 **RE:** AFED 2008, 18	**CM:** 3 **OD:** 9
				2010, 2014	**OD, CLF, SLF:** IDB-AIDIS-PAHO 2011, 132 **RE:** Costa Rica, Division of Operational and Evaluative Inspection 2016, 2, 22, 23, and 26	**LF:** 11 **OD:** 9 (Waste from 9.1 percent of covered population is dumped.) **CLF:** Waste from 23.5 percent of covered population goes to CLF. **RE:** 13 **SLF:** 12 (Waste from 67.5 percent of covered population goes to SLF.)
			97.0	2005	AFED 2008, 18	
			2.2	2015	Eurostat 2017	**CM:** 2 **UA:** 14
			17.6	2015	**OD, RE, UA:** Cuba ONEI 2016 **CLF:** Cuba, ONEI 2017 (population); Anon n.d. (waste disposed in landfills); Rebelde 2007 (generation rates)	**OD:** Calculated based on remainder of waste that was collected but not recycled or put in CLF in Havana. **CLF:** Calculated based on estimate of total waste received at unengineered landfills in Havana Province only divided by MSW generated nationwide; no information available on landfills outside of Havana; MSW generation estimates do not include bulky waste, industrial, or medical waste; also likely does not include commercial, but this cannot be confirmed. **RE:** Calculated based on estimates of waste recycled or composted compared with total generation. **UA:** Calculated based on waste produced by population without collection service, primarily in rural areas.
			98.0	2013	Pricewaterhouse-Coopers Aruba 2014	
			7.6	2015	Eurostat 2017	
			0.0	2015	Eurostat 2017	**CM:** 2
			0.0	2015	Eurostat 2017	**CM:** 2
			6.0	2005	UNSD 2013	**CLF:** Estimated based on population with access to formal collection services; likely that actual value is higher; value is supported by World Bank site visits. **UA:** 6 (Actual figure may be lower.)
			19.1	2017, 2015	**OD, RE:** Dominican Republic, Ministry of Environment and Natural Resources 2014 **SLF:** Dominican Republic, Ministry of Environment and Natural Resources and Ministry of Economy 2017	**OD:** Calculated based on the sum of other treatment and disposal options subtracted from total waste generated. **SLF:** 1 (Estimated based on the amount of waste taken to one sanitary landfill in Las Placetas, San Jose de las Matas.) **RE:** Includes all exported recyclables (metals, paper, carton, plastics, and glass); does not include items recycled in country. **Other:** Calculated based on the amount of waste from HH without collection services compared with total waste generation.

(Table continues on next page)

Country or economy	Region	Income	Open dump	Landfill unspecified	Controlled landfill	Sanitary landfill	Recycling	Com-posting	Anaerobic digestion	Incin-eration
Ecuador	LAC	UMIC	22.3		-	53.2	12.9			
Egypt, Arab Rep.	MENA	LMIC	84.0	7.0			12.5	7.0		
El Salvador	LAC	LMIC	13.8			78.2				
Estonia	ECA	HIC			7.4		24.7	3.6		51.4
Ethiopia	SSA	LIC	43.0							
Faeroe Islands	ECA	HIC					67.0			
Fiji	EAP	UMIC		52.0			5.5			
Finland	ECA	HIC		11.5			28.1	12.5		47.9
France	ECA	HIC		25.8			22.3	17.3		34.7
French Polynesia	EAP	HIC					39.0			
Germany	ECA	HIC		0.2			47.8	18.2		31.7
Greece	ECA	HIC		80.0			19.0			
Greenland	ECA	HIC		60.0						40.0

Advanced thermal treatment	Water-ways	Other	Un-accounted for	Year(s)	Source	Comment
			11.6	2015	Ecuador, Ministry of Environment 2018	1 **OD:** Calculated assuming that 29.5 percent of all collected waste (88.4 percent) that is not recycled is dumped. **SLF:** Calculated based on percentage of people with collection service for this disposal method times the percent of total collection coverage. **RE:** 14 **UA:** 6
				2013	GIZ and SWEEP-Net 2014b	**OD:** Reported as 80–88 percent in source (average used). **RE:** Reported as 10–15 percent in source (average used).
		7.9	0.1	2010	IDB-AIDIS-PAHO 2011, 132	**Other:** Includes open burning (7.3 percent) and waste disposed as cattle feed, dumped in WW, and so on.
			12.9	2015	Eurostat 2017	**CM:** 2
			57.0	2011	Global Methane Initiative 2011	
			33.0	2012	**RE:** Nordic Competition Authorities 2016, 59 **IN:** Frane, Stenmarck, and Gislason 2014	**RE:** Recovery includes incineration with recovery, CM, AD, RE, other recovery, and hazardous materials exported for treatment; mineral waste that is inert is usually landfilled or used for land reclamation. **IN:** Some incineration occurs but exact percentage unknown.
			42.6	2011, 2013	**LF:** Fiji, Department of Environment 2011 **RE:** Patriamby and Tanaka 2014, 274	**LF:** Calculated based on the amount of waste landfilled or dumped in 2010 compared with the amount of waste generated in 2011. **RE:** Average recycling rate derived from the recycling rates of Lautoka City (8.1 percent) and Nadi Town (2.8 percent).
			0.0	2015	Eurostat 2017	**CM:** 2
			0.0	2015	Eurostat 2017	
			61.0	2013	SPREP 2016	**OD:** 9 **LF:** 11 **SLF:** 12 (99 waste disposal sites, of which 5 are SLF, 3 are controlled dumps, 8 are authorized open dumps, and 80 are temporary unregulated dumps.) **CM:** 3 (1 large-scale compost program in Tahiti.)
			2.0	2015	Eurostat 2017	
		1.0		2014, 2011	Greece, Ministry of Environment and Energy 2015, 17	**LF:** Reported as disposed of. **RE:** Reported as recovered (recycling and composting). **Other:** Reported as unregistered management.
				2010	Eisted and Christensen 2011	**LF:** Calculated based on the amount of waste landfilled compared with amount generated (average used). **IN:** Calculated based on the amount of waste incinerated compared with amount generated.

(Table continues on next page)

Country or economy	Region	Income	Open dump	Landfill unspecified	Controlled landfill	Sanitary landfill	Recycling	Com- posting	Anaerobic digestion	Incin- eration
Grenada	LAC	UMIC			98.3			0.2		
Guam	EAP	HIC				64.0	17.9			
Guatemala	LAC	LMIC	69.8		9.6	15.4				
Guinea	SSA	LIC					5.0			
Guyana	LAC	UMIC		61.4			0.5			
Haiti	LAC	LIC			9.9					
Honduras	LAC	LMIC	15.0		59.9	11.3				
Hong Kong SAR, China	EAP	HIC		66.0			34.0			
Hungary	ECA	HIC		53.6			25.9	6.2		14.1
Iceland	ECA	HIC		30.3			55.8	2.9		1.9
India	SAR	LMIC	77.0				5.0	18.0		

Advanced thermal treatment	Water-ways	Other	Un-accounted for	Year(s)	Source	Comment
		1.5		2011	Grenada, Population and Housing Census 2011, 35	**CLF:** The two landfills are controlled landfills; they are being upgraded as part of a Caribbean Development Bank project. **CM:** Represents an approximate value of waste composted in HH. **Other:** Includes open burning (0.7 percent), dumping (0.2 percent), dumping on land (0.4 percent), burying (0.1 percent), and other unspecified (0.1 percent).
			18.2	2012, 2011	**SLF, RE:** Guam 2013 **CM:** SPREP 2016	**SLF:** Calculated based on amount disposed of in Layon Landfill and amount of waste generated.
			5.2	2010	IDB-AIDIS-PAHO 2011, 132	**UA:** 15
			95.0	2005	AFED 2008, 18	
			38.1	2011, 2010	Guyana, Ministry of Communities n.d.	**UA:** Refers to remainder of waste not disposed of in landfill or recycled, which is mainly disposed of in CLF and ODs; a small portion is recycled through glass and scrap metal recycling programs.
			90.1	2012	IHSI, IRD, Dial, Nopoor, ANR 2014 (coverage); IHSI 2015 (population); SWANA Haiti Response Team 2010 (generation - Port au Prince); Naquin 2016 (generation per capita urban and rural areas – Cap-Haïtien)	**CLF:** Calculated assuming that total waste collected in the metropolitan area of Port au Prince is disposed of in the Trutier landfill as a percentage of the total waste generated countrywide. **UA:** Includes all waste collected from other urban and rural areas.
			13.8	2010	IDB-AIDIS-PAHO 2011, 132	**UA:** 15 and all other disposal methods.
				2016	Hong Kong, Environmental Protection Department, Statistics Unit 2017	
			0.1	2015	Eurostat 2017	
		9.1		2013	Iceland, Statistics Iceland 2015, 429	Calculated based on actual values provided. **Other:** Includes other recovery (8.38 percent), other disposal (0.38 percent), and hazardous waste exported for treatment (0.38 percent).
				2016, 2013	**OD, CM:** India CPCB 2017 **RE:** Mahapatra 2013	1 **OD:** Assuming 100 percent of rural waste is dumped; 77.96 percent of urban waste is dumped based on CPCB data, less amount recycled. **RE:** Based on estimate that 15,342 tonnes of plastic is disposed of every day, 60 percent of which is recycled. **CM:** Value refers to total amount of waste processed from composting, RDF, and biogas.

(Table continues on next page)

Country or economy	Region	Income	Open dump	Landfill unspecified	Controlled landfill	Sanitary landfill	Recycling	Com- posting	Anaerobic digestion	Incin- eration
Indonesia	EAP	LMIC	10.0	69.0			7.0			
Iran, Islamic Rep.	MENA	UMIC	72.0		10.0		5.0	12.0	0.3	0.4
Iraq	MENA	UMIC	100.0							
Ireland	ECA	HIC			41.0		33.0	6.0		17.0
Isle of Man	ECA	HIC		25.0			50.0			25.0
Israel	MENA	HIC		75.0			25.0			
Italy	ECA	HIC		26.5			25.9	17.6		19.0
Jamaica	LAC	UMIC			64.0					
Japan	EAP	HIC			1.1		4.9	0.4	0.1	80.2
Jordan	MENA	LMIC	45.0	48.0			7.0			
Kazakhstan	ECA	UMIC	60.1				2.9			
Kenya	SSA	LMIC					8.0			
Korea, Rep.	EAP	HIC		16.0			58.0	1.0		25.0
Kosovo	ECA	LMIC	33.6		66.4					

Advanced thermal treatment	Water-ways	Other	Un-accounted for	Year(s)	Source	Comment
		14.0		2016	Damanhuri 2017, 3	**RE:** 13 **OD:** Referred to as "illegal dumping" in source. **Other:** Includes disposal in rivers, on streets, gardens, and so on (9 percent) and open burning (5 percent).
			0.3	2017	Abedini 2017	1 **RE:** Value for material segregated in sorting plants. **AD:** There is only one facility in Tehran with capacity of 150 tonnes/day. **IN:** There is only one facility in Tehran with capacity of 200 tonnes/day.
				2015	Iraq, Ministry of Environment 2015	
		2.0	1.0	2012	Ireland, EPA 2014, 1	**CM:** 2 **Other:** 34 percent of MSW managed in Ireland was exported for energy recovery and recycling.
				2011	Isle of Man, Department of Infrastructure n.d., 12	
				2017	Israel, Ministry of Environmental Protection 2017	**LF, RE:** Reported as, "Some 75 percent of the waste in the country is buried in landfills while only about 25 percent is recycled."
			11.0	2015	Eurostat 2017	**CM:** 2 **UA:** 14
		29.0	7.0	2016, 2011	**CLF:** Jamaica NSWMA 2016 **Other:** Jamaica 2011	**CLF:** 14 **Other:** Calculated based on the amount treated or disposed of compared with the amount generated [burned (34.58 percent), buried (0.60 percent), and WW (0.82 percent)].
		13.3		2015	Japan, Ministry of the Environment 2015	**SLF:** 12 (2 out of 1,718 facilities.) **IN:** Includes ATT.
				2014	GIZ and SWEEP-Net 2014c, 7	
			37.0	2012	World Bank n.d., 42	**OD:** Calculated based on the amount of waste disposed of in dumpsites or landfills compared with the amount of MSW generated. **RE:** Calculated based on the amount of waste recycled and processed compared with the amount of MSW generated.
			92.0	2009	UNECA 2009, 24	8 **RE:** 13
				2014	OECD 2017	
				2010	Kosovo, Ministry of Environment and Spatial Planning 2013, 27	

(Table continues on next page)

Country or economy	Region	Income	Open dump	Landfill unspecified	Controlled landfill	Sanitary landfill	Recycling	Composting	Anaerobic digestion	Incineration
Kuwait	MENA	HIC	100.0							
Kyrgyz Republic	ECA	LMIC	100.0							
Lao PDR	EAP	LMIC	60.0		30.0		10.0			
Latvia	ECA	HIC		57.6			21.2	5.5		
Lebanon	MENA	UMIC	29.0		48.0		8.0	15.0		
Liechtenstein	ECA	HIC					64.6			
Lithuania	ECA	HIC		54.0			22.9	10.2		11.5
Luxembourg	ECA	HIC		17.7			28.4	19.7		34.0
Macao SAR, China	EAP	HIC					20.0			
Macedonia, FYR	ECA	UMIC		99.7			0.2	0.1		
Madagascar	SSA	LIC	96.7					3.5		
Malaysia	EAP	UMIC		71.5		10.0	17.5	1.0		
Maldives	SAR	UMIC					7.0			6.0
Malta	MENA	HIC				89.6	6.7			0.4
Marshall Islands	EAP	UMIC					30.8	6.0		
Mauritania	SSA	LMIC	54.7		37.3		8.0			
Mauritius	SSA	UMIC				91.0		9.0		
Mexico	LAC	UMIC	21.0			74.5	5.0			
Moldova	ECA	LMIC	35.1				15.3			

Advanced thermal treatment	Water-ways	Other	Un-accounted for	Year(s)	Source	Comment
				2014	Alsulaili et al. 2014	
				2010	Barieva 2012	Calculated based on amount of domestic waste disposed of compared with total amount of domestic waste generated.
				2015	**CLF:** Keohanam 2017 **RE:** CCAC n.d.(b)	
			15.7	2015	Eurostat 2017	**CM:** 2
				2014	GIZ and SWEEP-Net 2014d, 8	
			35.4	2015	**RE:** Liechtenstein, Office of Statistics 2018, 7	**RE:** Value is for urban waste.
			1.4	2015	Eurostat 2017	
			0.3	2015	Eurostat 2017	**CM:** 2
			80.0	2014	**RE:** Macao SAR, China 2014 **IN:** Macao SAR China, DSEC 2017	**RE:** Includes plastics, rubber, paper, metal, and other recoverable waste; approximated from figure in source. **IN:** Some incineration occurs at Macao Refuse Incineration Plant, which treats domestic and ICI waste, but exact percentage unknown.
				2013	**LF, RE, CM:** Macedonia, FYR, Ministry of Environment and Physical Planning 2014, 92 **IN:** Dimishkovska and Dimishkovski 2012, 264	
				2007	UNSD 2016	
				2017, 2016	UNCRD 2017	
	63.0	24.0		2016	Maldives, MEE 2017, 173	All values are specifically for kitchen waste disposal. **Other:** Includes bury (17 percent) and open burning (7 percent).
			3.4	2015	Eurostat 2017	**SLF:** Uncontrolled landfills were replaced with two major engineered landfills in 2004 and 2006.
			63.2	2007	UNSD 2016	
				2009	GIZ and SWEEP-Net 2010a	
				2012	UNSD 2016	
				2013	Mexico, SEMARNAT 2016, 444–45	**AD:** 16 **IN:** Only for hazardous waste and health care waste.
			49.6	2015	Moldova, Statistica Moldovei 2016, 58	**OD:** Calculated based on the amount deposited compared with the amount of waste generated. **RE:** Calculated based on the amount recycled compared with the amount of waste generated.

(Table continues on next page)

Country or economy	Region	Income	Open dump	Landfill unspecified	Controlled landfill	Sanitary landfill	Recycling	Com-posting	Anaerobic digestion	Incin-eration
Monaco	ECA	HIC					5.4			85.0
Mongolia	EAP	LMIC	93.5							
Montenegro	ECA	UMIC	91.6				5.4			
Morocco	MENA	LMIC	52.0			37.0	8.0	1.0		
Mozambique	SSA	LIC	99.0				1.0			
Myanmar	EAP	LMIC								
Namibia	SSA	UMIC					4.5			
Nepal	SAR	LIC		37.0				2.9		
Netherlands	ECA	HIC		1.4			24.6	27.1		47.4
New Zealand	EAP	HIC		100.0						
Nicaragua	LAC	LMIC	59.3	19.6						
Niger	SSA	LIC	64.0	-			4.0			-
Nigeria	SSA	LMIC		40.0						
Northern Mariana Islands	EAP	HIC					36.0			
Norway	ECA	HIC				3.4	26.2	16.7		52.4
Oman	MENA	HIC	100.0		0.0					
Pakistan	SAR	LMIC	50.0	40.0			8.0	2.0		
Panama	LAC	UMIC	23.4		16.0	41.7				
Papua New Guinea	EAP	LMIC			62.0		2.0			
Paraguay	LAC	UMIC	23.4		40.2	36.4				
Peru	LAC	UMIC	56.4		15.6	24.0	4.0			

Advanced thermal treatment	Water-ways	Other	Un-accounted for	Year(s)	Source	Comment
			9.6	2012, 2013	RE: UNSD 2016 IN: Monaco, Directorate of Environment 2013	IN: Calculated based on 39,000 tonnes from the principality (including sewage sludge) that is incinerated; actual incineration rate is higher as waste is imported.
			6.5	2016	Delgerbayar 2016	
			3.0	2016	OD, RE, UA: Eurostat 2017 SLF: ZWMNE 2016	SLF: 12 [2 SLF in Podgorica (Livade) and Mozura.]
		2.0		2014	GIZ and SWEEP-Net 2014e, 7	CM: Value is given as <1 percent in source.
				2014	Tas and Belon 2014	RE: < 1 percent of waste recycled (estimated); waste that is not recycled is either dumped or buried.
		8.0	92.0	2010	Thein 2010	OD: 9 Other: Value refers to open burning.
			95.5	2005	AFED 2008, 18	
			60.1	2013	ADB 2013	LF: Source says disposed of in sanitary landfills, but not in a sanitary manner. CM: Value for all composting not known. UA: Value represents uncollected waste.
				2015	Eurostat 2017	CM: 2
				2015	UNSD 2016	
		21.1		2010	IDB-AIDIS-PAHO 2011, 132	Other: Includes open burning (7.5 percent).
		12.0	20.0	2005	UNSD 2016	Other: Refers to open burning.
		60.0		1995	OD, RE: Ayuba et al. 2013 LF, Other: IPCC 2006, 17	RE: 4 OD: 9
			64.0	2016	US EPA 2016	
			1.4	2015	OECD 2017	
				2017	Ouda 2017	1; 8
				2017	Korai, Mahar, and Uqaili 2017, 348	
		18.9		2010	IDB-AIDIS-PAHO 2011, 132	Other: 15 and open burning (4.7 percent).
		37.0		2016	Papua New Guinea, NCDC 2016, 45	RE: Recycling is limited to cans, plastic, glass containers, and food for piggeries. Other: Includes illegal dumping and open burning.
				2010	IDB-AIDIS-PAHO 2011, 132	
				2014, 2012	OD, CLF, SLF: Peru, Ministry of Environment 2016, 21 RE: Peru, Ministry of Environment 2013, 3	OD: Waste not disposed of in SLF was disposed of inadequately. SLF: 14

(Table continues on next page)

Country or economy	Region	Income	Open dump	Landfill unspecified	Controlled landfill	Sanitary landfill	Recycling	Com- posting	Anaerobic digestion	Incin- eration
Philippines	EAP	LMIC					28.0			
Poland	ECA	HIC		44.3			26.4	16.1		13.2
Portugal	ECA	HIC		49.0			16.2	14.1		20.7
Puerto Rico	LAC	HIC		66.5			14.0			
Qatar	MENA	HIC		93.0			3.0			4.0
Romania	ECA	UMIC		72.0			5.7	7.5		2.4
Russian Federation	ECA	UMIC	95.0				4.5			
Samoa	EAP	UMIC			31.0		36.0			
San Marino	ECA	HIC					45.1			
Saudi Arabia	MENA	HIC			85.0		15.0			
Senegal	SSA	LIC	43.8		5.1					
Serbia	ECA	UMIC		73.9			0.8			
Singapore	EAP	HIC		2.0			61.0			37.0
Slovak Republic	ECA	HIC		68.7			7.6	7.3		10.7
Slovenia	ECA	HIC		22.7			46.4	7.7		17.1
Solomon Islands	EAP	LMIC								
South Africa	SSA	UMIC			72.0		28.0			

Advanced thermal treatment	Water-ways	Other	Un-accounted for	Year(s)	Source	Comment
			72.0	2014	Modak et al. 2017, 235	
				2015	Eurostat 2017	**CM:** 2
				2014	Eurostat 2017	**CM:** 2
			19.5	2007, 2013	**LF:** Energy Answers 2012, 2 **RE:** Pricewaterhouse-Coopers Aruba 2014	**LF:** 14 (Assumed disposed of in landfill, as there are 29 operating landfills.)
				2014	Ayoub, Musharavati, and Gabbar 2014, 96	
			12.5	2015	Eurostat 2017	**LF:** Refers to waste disposal in general as there is no information on the type of disposal; in Romania there is a combination of controlled and sanitary landfills with landfill gas recovery. **UA:** 14
			0.5	2012	**OD:** IFC 2012, 5 **RE:** Russia, Ministry of Natural Resources and Ecology 2012, 7	**RE:** Source provides a range of 4–5 percent (average used).
			33.0	2013	SPREP 2016	**CLF:** 14 [Semi-aerobic landfill (Fukuoka method).] **RE:** Refers to amount exported or recycled or reused locally.
			55.0	2016	San Marino, AASS 2016	**RE:** Value refers to amount of waste that is collected separately; all of this waste is recovered in some form.
				2015	Saudi Arabia n.d.	**CLF:** Value based on personal knowledge and the difference between total disposal (100 percent) and amount recycled (15 percent). **RE:** Includes recycling and treatment.
		4.7	46.4	2014	**OD, CLF, Other:** ANSD 2014 **RE, CM:** Gret-LVIA-Pacte 2006	**OD:** Includes dumping (42.2 percent) and informal burial (1.6 percent). **RE:** Most households engage in recycling activities; there are various societies devoted to the recycling of plastic (PROPLAST), paper (PRONAT), and aluminum (SELMEG). **Other:** Includes open burning (3.5 percent).
			25.3	2015	**OD:** Anthouli et al. 2013, 27 **LF, RE:** Eurostat 2017	**OD:** There are 3,582 identified landfills, of which 165 are municipality landfills, 5 are SLF, and the rest are dumps.
				2015	Singapore, Ministry of the Environment and Water Resources 2017	
			5.7	2015	Eurostat 2017	
			6.2	2015	Eurostat 2017	**CM:** 2
81.0			19.0	2015	Solomon Islands, MECDM 2015	8
				2011	South Africa, Department of Environmental Affairs 2012	**CLF, RE:** Include MSW and C&I waste; excludes C&D, hazardous, and inert waste.

(Table continues on next page)

Country or economy	Region	Income	Open dump	Landfill unspecified	Controlled landfill	Sanitary landfill	Recycling	Com-posting	Anaerobic digestion	Incin-eration
Spain	ECA	HIC		55.1			16.8	16.5		11.6
Sri Lanka	SAR	LMIC	85.0				12.8	5.0		
St. Kitts and Nevis	LAC	HIC			100.0					
St. Lucia	LAC	UMIC			96.8			0.1		
St. Vincent and the Grenadines	LAC	UMIC		99.9				0.1		
Sudan	SSA	LMIC		82.0						
Suriname	LAC	UMIC	63.0							
Sweden	ECA	HIC		0.8			32.4	15.6		51.2
Switzerland	ECA	HIC					32.0	21.0		47.0
Syrian Arab Republic	MENA	LMIC	80.0	20.0			2.5	1.5		
Taiwan, China	EAP	HIC		34.8						64.2

Advanced thermal treatment	Water-ways	Other	Un-accounted for	Year(s)	Source	Comment
				2015	Eurostat 2017	**CM:** 2
				2016	Sri Lanka, Ministry of Mahaweli Development and Environment 2016	**OD:** Reported as more than 85 percent of waste dumped unscientifically.
				2017	SIDS DOCK 2015, 14	
		2.6	0.5	2010	St. Lucia Government Statistics Department 2011	**Other:** Includes open burning (1.5 percent); dumping on land (0.6 percent); dumping in river, sea, or pond (0.4 percent); and burying (0.1 percent). **UA:** Includes other (0.2 percent) and not stated (0.3 percent).
		3.6		2012	**LF, CM:** St. Vincent and the Grenadines, Statistical Office 2012, 45 **Other:** St. Vincent and the Grenadines, Statistical Office n.d.	**LF:** Five landfills are operational; MSW either is sent to landfills or composted. **CM:** Includes that which is composted after collection and at HH level (0.1 percent of households compost as their major form of disposal). **Other:** Includes burning (2.6 percent); burial (0.2 percent); open dumping (0.4 percent); dumping in river, sea, or pond (0.2 percent); and other not specified (0.2 percent); values are for percentage of HH undertaking waste disposal.
	18.0			2003	IPCC 2006, 17	
			37.0	2013	Viren 2013	**OD:** Open dumping is the main waste disposal method for the country; the largest dump that is most similar to a formal landfill still has fires, leachate management deficiencies, and animals on site; formally collected waste is sent to a dump. **UA:** 6, which is managed by households in a variety of ways, but is most likely dumped.
				2015	Eurostat 2017	**CM:** 2
				2015	OECD 2017	**LF:** There are no landfills for MSW, but they exist for inert materials, stabilized residues, and bioreactor landfills.
				2010	GIZ and SWEEP-Net 2010b	**OD:** Source provides a value of about 80 percent. **LF:** Source provides a value of about 20 percent landfilled. **RE:** Source provides a range of 2–3 percent (average used). **CM:** Source provides a range of 1–2 percent (average used).
		1.0		2002	Tsai and Chou 2006	**IN:** Primary method of disposal. **Other:** Value includes composting and dumping.

(Table continues on next page)

Country or economy	Region	Income	Open dump	Landfill unspecified	Controlled landfill	Sanitary landfill	Recycling	Com-posting	Anaerobic digestion	Incin-eration
Tajikistan	ECA	LMIC	100.0							
Tanzania	SSA	LIC	69.0							
Thailand	EAP	UMIC	53.5	27.0			19.1			0.4
Togo	SSA	LIC	96.2				2.0	1.8		
Tonga	EAP	UMIC		40.0						
Trinidad and Tobago	LAC	HIC	84.0		12.0			0.8		
Tunisia	MENA	LMIC	21.0	70.0			4.0	5.0		
Turkey	ECA	UMIC	44.0			54.0		1.0		
Turkmenistan	ECA	UMIC		100.0						
Tuvalu	EAP	UMIC	14.5				15.0			
Uganda	SSA	LIC	87.0				7.0	6.0		

Advanced thermal treatment	Water- ways	Other	Un- accounted for	Year(s)	Source	Comment
				2015	OD: Boboeva 2015, 2 RE: UNECE 2017	8 OD: MSW is neither sorted nor treated; uncontrolled dumping is widespread. RE: The country generally lacks recycling infrastructure, except for scrap metals and paper; collection of waste paper, glass, and other recyclables is primarily done by the informal sector.
		22.6	8.5	2012	Tanzania, NBS and OCGS 2014	OD: Includes informally disposed of in pits or buried (36.2 percent). Other: Value refers to open burning. UA: Amount is collected by company or authority but disposal mechanism is unspecified.
				2012	Intharathirat and Salam 2015, 35	OD: Reported as "disposed improperly." LF: Reported as "disposed properly."
			-	2014, 2012	OD: CCAC n.d.(a) RE, CM: UNSD 2016	OD: All waste is disposed of in an "open landfill" (dump) that is not sorted and precollected.
			60.0	2012	ADB 2014	Calculated based on the amount collected compared with the amount generated; all collected waste is landfilled.
			3.2	2011	Trinidad and Tobago, EMA n.d.	All values calculated based on National Census 2011. CM: 1.17 percent of HH waste is composted by HH and HH waste makes up two-thirds of all waste generation. UA: 4.5 percent of HH waste is not collected or composted; includes waterways. (0.1 percent) and burning (2.2 percent).
				2014	GIZ and SWEEP-Net 2014g	
			1.0	2015	OD, SLF, UA: Bakas and Milios 2013, 5 CM: OECD 2017	UA: Consists of biological treatment or disposal by other methods.
				2013	Zoï Environment Network 2013, 25	LF: Reported as "almost all" going to landfills.
			70.5	2013	SPREP 2016	CM: 3 OD: Calculated based on the amount landfilled or dumped compared with the amount of waste generated; there are 9 authorized dumps in Tuvalu.
				2017	OD, RE, SLF: KCCA-IFC 2017 CM: Okot-Okumu 2012, 7	RE: 4 OD, SLF, RE: Estimate based on total waste generated and report that Kampala has the only SLF receiving 1,300 tonnes/day; an estimated 6 percent of waste is removed from the waste stream for recycling; waste in other areas is dumped. CM: 8 (Composting is being practiced in more than 11 urban councils of Uganda under the Clean Development Mechanism under the Kyoto Protocol pilot project promoted by the World Bank, but no actual values available.)

(Table continues on next page)

Country or economy	Region	Income	Open dump	Landfill unspecified	Controlled landfill	Sanitary landfill	Recycling	Com-posting	Anaerobic digestion	Incin-eration
Ukraine	ECA	LMIC		94.1			3.2			2.7
United Arab Emirates	MENA	HIC	62.0	9.0			20.0	9.0		
United Kingdom	ECA	HIC		22.6			27.3	16.2		31.4
United States	NA	HIC		52.6			34.6			12.8
Uruguay	LAC	HIC	17.5		61.7	10.5	8.0			
Uzbekistan	ECA	LMIC	60.0							
Vanuatu	EAP	LMIC	11.3				37.0			
Vietnam	EAP	LMIC					23.0	15.0		
West Bank and Gaza	MENA	LMIC	67.0		33.0		0.5	0.5		
Yemen, Rep.	MENA	LMIC	25.0		12.0		8.0			
Zimbabwe	SSA	LIC					16.0			

Note: Year refers to year of data, unless otherwise specified.

AD = anaerobic digestion; ATT = advanced thermal treatment; C&D = construction and demolition; C&I = commercial and institutional; CLF = controlled landfill; CM = composting; CPCB = Central Pollution Control Board (Government of India); EAP = East Asia and the Pacific; ECA = Europe and Central Asia; HH = household; HIC = high-income country; ICI = institutional, commercial, and industrial; IN = incineration; LAC = Latin America and the Caribbean; LF = landfill unspecified; LIC = low-income country; LMIC = lower-middle-income country; MENA = Middle East and North Africa; MSW = municipal solid waste; NA = North America; OD = open dumping; RDF = refuse-derived fuel; RE = recycling; SAR = South Asia; SLF = sanitary landfill; SSA = Sub-Saharan Africa; UA = unaccounted for; UMIC = upper-middle-income country; WW = waterways.

1. Personal communication.
2. Value includes composting and anaerobic digestion.
3. Composting occurs but exact percentage is unknown.
4. Recycling occurs but exact percentage is unknown.
5. Value for MSW only.
6. Value refers to uncollected waste.
7. Anaerobic digestion occurs but exact percentage is unknown.
8. Year refers to year of publication.
9. Open dumping occurs but exact percentage is unknown.
10. Open burning occurs but exact percentage is unknown.
11. Some landfilling occurs but exact percentage is unknown.
12. Some sanitary landfilling occurs but exact percentage is unknown.
13. Value includes recycling and composting.
14. Calculated based on the amount treated or disposed of compared with the amount generated, which is reported in appendix A.
15. Value includes dumping in waterways and usage as animal feed.
16. According to source, being established but exact status unknown.

Advanced thermal treatment	Water-ways	Other	Un-accounted for	Year(s)	Source	Comment
				2015	Business Sweden, The Swedish Trade and Invest Council 2016, 5	**LF:** 6,000 landfills, of which 31 percent are not certified or licensed.
				2015	Abu Dhabi SCAD 2016	**OD:** 5; reported as dumpsite and other; all values for Abu Dhabi emirate only.
			2.6	2015	Eurostat 2017	**CM:** 2
				2014	US EPA 2014	**RE:** 13
			2.3	2013, 2011	**OD, CLF, SLF:** CSI Ingenieros 2011; Anon n.d.(d); LKSur 2013, 8 **RE:** Oriental Republic of Uruguay 2004, 9; CSI Ingenieros 2011	**OD, CLF, SLF:** Calculated based on amount disposed of at type of facility and coverage rate. **RE:** Calculated based on values available from formal recycling programs; does not include informal recycling or other formal recycling activities.
			40.0	2011	CER 2011, 28	**OD:** According to the State Committee for Nature Protection, there are 178 registered dumps and several hundred additional unregistered dumps.
			51.7	2013	SPREP 2016	Calculated based on the amount landfilled or dumped compared with the amount of waste generated.
			62.0	2014, 2013	**RE:** Patriamby and Tanaka 2014, 364 **CM:** Vietnam WENID 2013	**RE:** Reported as a range of 18–28 percent (average used).
				2013	GIZ and SWEEP-Net 2014f	**RE, CM:** Value in source given as < 0.5 percent.
	8.0		47.0	2016	Al-Eryani 2017	1
			84.0	2005	AFED 2008, 18	

References

Abedini, Ali R. 2017. Solid waste management specialist, and founder and CEO, ISWM Consulting Ltd. Personal communication between A. Abedini and Iran Municipal and Rural Management Organization (MRMO).

ABRELPE (Brazilian Association of Public Cleaning and Special Waste Companies). 2015. "Overview of Solid Waste in Brazil 2015" ["Panorama dos Resíduos Sólidos no Brasil 2015"]. ABRELPE, São Paulo.

Abu Dhabi SCAD (Statistics Centre Abu Dhabi). 2016. "Waste Statistics 2015." Government of Abu Dhabi.

ADB (Asian Development Bank). 2011. *Toward Sustainable Municipal Organic Waste Management in South Asia - A Guidebook for Policy Makers and Practitioners*. Manila: ADB.

————. 2013. *Solid Waste Management in Nepal: Current Status and Policy Recommendations*. Mandaluyong City, Philippines: ADB.

————. 2014. "Solid Waste Management in the Pacific: Tonga Country Snapshot." ADB Publication Stock No. ARM146616-2, ADB, Manila. https://www.adb.org/sites/default/files/publication/42660 /solid-waste-management-tonga.pdf.

AFED (Arab Forum for Environment and Development). 2008. "Arab Environment: Future Challenges." Edited by Mostafa K. Tolba and Najib W. Saab. Arab Forum for Environment and Development, Beirut.

Al-Eryani, Muammer. 2017. Director of Solid Waste Management, Ministry of Local Administration (MoLA), Government of the Republic of Yemen. Personal communication with the World Bank.

Al Sabbagh, Maram K., Costas A. Velis, David C. Wilson, and Christopher R. Cheeseman. 2012. "Resource Management Performance in Bahrain: A Systematic Analysis of Municipal Waste Management, Secondary Material Flows and Organizational Aspects." *Waste Management and Research* 30 (8): 813–24. http://journals.sagepub.com/doi/pdf/10.1177 /0734242X12441962.

Alsulaili, Abdalrahman, Bazza Al Sager, Hessa Albanwan, Aisha Almeer, and Latifa Al Essa. 2014. "An Integrated Solid Waste Management System in Kuwait." 5th International Conference on Environmental Science and Technology. IPCBEE vol. 69. IACSIT Press, Singapore. doi:10.7763/IPCBEE. V69. 12.

Amec Foster Wheeler. 2016. "National Solid Waste Management Strategy for the Cayman Islands: Final Report." Amec Foster Wheeler Environment and Infrastructure UK Limited, for the Cayman Islands Government. http://ministryofhealth.gov.ky/sites/default/files/36082%20Strategy %20Final%20Report%2016229i1.pdf.

Anon. n.d. (a). Joint study by representatives of the following organizations: Departamento de Estudios sobre Contaminación Ambiental, Centro Nacional de Investigaciones Científicas, Consultoría de Ingeniería y

Arquitectura, Centro de Estudios de Ingeniería de Procesos, Instituto Superior Politécnico "José Antonio Echeverría." Laboratorio de Análisis de Residuos (LARE), Dirección Provincial de Servicios Comunales, Ministerio de Economía y Planificación, Estimated Entries to Campo Florido. Cited in Periodísmo de Barrio article, but original source unknown.

Anon. n.d. (b). "Sanitary Filling of Roses" ["RELLENO SANITARIO DE LAS ROSAS"]. Maldonado; Seminar of Small Municipalities, CEMPRE.

ANSD (National Agency of Statistics and Demography). 2014. "General Census of Population and Housing, Agriculture and Livestock. Final Report" ["Recensement Général de la Population et de l'Habitat, de l'Agriculture et de l'Elevage. RAPPORT DEFINITIF"]. ANSD, Government of Senegal.

Anthouli, Aida, Konstantine Aravossis, Rozy Charitopoulou, Bojana Tot, and Goran Vujic. 2013. "Opportunities and Barriers of Recycling in Balkan Countries: The Cases of Greece and Serbia." Hellenic Solid Waste Management Association and Serbian Solid Waste Management Association, with the International Solid Waste Association.

Antigua and Barbuda, Statistics Divison. 2014. "Antigua and Barbuda 2011 Population and Housing Census–Book of Statistical Tables." April.

Argentina SIDSA (System of Indicators of Sustainable Development). 2015. "System of Indicators of Sustainable Development, Argentina." 8th Edition. ["Sistema de Indicadores de Desarrollo Sostenible, Argentina." Octava Edición.] SIDSA.

Armel, Lucien. 2017. AWAH [Manga Solid Waste Disposal among Urban Agricultural Households in Lowland Area of Yaounde]. Third International Scientific Symposium "Agrosym Jahorina 2012."

Australia, Department of the Environment and Energy. 2017. "Australian National Waste Report 2016." Department of the Environment and Energy and Blue Environment Pty Ltd.

Ayoub, Nasser, Farayi Musharavati, and Hossam A. Gabbar. 2014. "A Future Prospect for Domestic Waste Management in Qatar." International Conference on Earth, Environment and Life Sciences (EELS-2014), Dubai, December 23–24. http://iicbe.org/upload/9363 C1214080.pdf.

Ayuba, Kadafa Adati, Latifah Abd Manaf, Abdullah Ho Sabrina, and Sulaiman Wan Nur Azmin. 2013. "Current Status of Municipal Solid Waste Management Practise in FCT Abuja." *Research Journal of Environmental and Earth Sciences* 5 (6): 295–304.

Azerbaijan, Ministry of Economy. 2017. "National Solid Waste Management Strategy Plan: Volume I (Main Report)." Aim Texas Trading, LLC – ICP Joint Venture, on behalf of Ministry of Economy, Government of Azerbaijan, Baku.

Bakas, Ioannis, and Leonidas Milios. 2013. "Municipal Waste Management in Turkey." European Environment Agency, Copenhagen.

Barieva, A. 2012. "Waste Production and Disposal in the Kyrgyz Republic." Paper presented at the Waste Statistics Seminar, Geneva, April 11–13.

Belarus, Ministry of Housing and Utility. 2017a. Personal communication with the World Bank.

———. 2017b. "Report on Sanitation of Cities and Populated Areas for 2016."

Bhutan, National Environment Commission. 2016. "Bhutan State of Environment Report 2016." National Environment Commission, Bhutan.

Boboeva, Shahnoza. 2015. "Current State of Waste Management in Tajikistan and Potential for a Waste-to-Energy Plant in Khujand City." Master's thesis, Columbia University, Earth Engineering Center, New York.

Bolivia, MMAyA/VAPSB/DGGIRS. 2016. "National Infrastructure Program for Solid Waste Management, 2017–2020" ["Programa Nacional de Infraestructuras en Gestión Integral de Residuos Sólidos, 2017–2020"]. General Directorate of Integrated Solid Waste Management, Sub-Ministry of Potable Water and Basic Sanitation, Ministry of Environment and Water [Dirección General de Gestión Integral de Residuos Sólidos, Viceministerio de Agua Potable y Saneamiento Básico, Ministerio de Medio Ambiente y Agua], Government of Bolivia.

Bosnia and Herzegovina, BHAS (Agency for Statistics of Bosnia and Herzegovina). 2016. "First Release: Environment—Public Transportation and Disposal of Municipal Waste." BHAS, Bosnia and Herzegovina, Sarajevo.

Brazil SNIS (National Sanitation Information System). 2017. "Diagnosis of Urban Solid Waste Management - 2015" ["Diagnóstico do Manejo de Resíduos Sólidos Urbanos – 2015"]. SNIS, Ministry of Cities, National Secretariat of Environmental Sanitation.

Business Sweden, The Swedish Trade and Invest Council. 2016. "Solid Waste Management in Ukraine—Market Insights." Business Sweden in Ukraine, Kiev.

Canada, Statistics Canada. 2012. "Human Activity and the Environment: Waste Management in Canada." Environment Accounts and Statistics Division. Catalogue no. 16-201-X.

CCAC (Climate and Clean Air Coalition). n.d.(a). "Solid Waste Management City Profile: Lomé, Togo." Municipal Solid Waste Initiative. http://waste.ccac-knowledge.net.

———. n.d.(b). "Vientiane Capital, Lao People's Democratic Republic: Solid Waste Management City Profile." Accessed April 17, 2017. http://www.waste.ccacoalition.org/sites/default/files/files/vientiane-_city_profile_vientiane_capital_lao.pdf.

CER (Center for Economic Research). 2011. "Improvement of Urban Governance and Urban Infrastructure in Uzbekistan: Problems and the Search for New Mechanisms and Tools." [Центр экономических исследований. "Совершенствование городского управления и инфраструктуры городов в Узбекистане: проблемы и поиск новых механизмов и инструментов"]. Accessed April 18. http://www.unece .org/fileadmin/DAM/hlm/prgm/cph/experts/uzbekistan/UZB-Urban -Development-2011-RUS.pdf.

Chile, CONAMA (National Environmental Commission). 2010. "First Report on Solid Waste Management in Chile" ["Primer Reporte del Manejo de Residuos Sólidos en Chile"]. CONAMA, Government of Chile.

Cissé, Sidi Mahamadou. 2015. Director of sanitation, Government of Burkina Faso. Personal communication with the World Bank.

Costa Rica, Division of Operational and Evaluative Inspection. 2016. "Operational Audit Report on Municipalities Management to Guarantee the Efficient Provision of the Ordinary Waste Collection Service" ["Informe De Auditoría Operativa Acerca De La Gestión De Las Municipalidades Para Garantizar La Prestación Eficaz Y Eficiente Del Servicio De Recolección De Residuos Ordinarios"].

CSI Ingenieros. 2011. "Background Information for the Design of a Solid Waste Strategic Plan" ["Informacion de Base para el Diseno de Un Plan Estrategico de Residuos Solidos"]. CSI Ingenieros. Pittamiglio Studio [Estudio Pittamiglio], Montevideo, Uruguay.

Cuba, ONEI (National Office of Statistics and Information). 2016. "Statistical Yearbook of Cuba 2015" ["Anuario Estadistico de Cuba 2015"]. ONEI, Government of Cuba.

———. 2017. "Catalogue of Publications."

Damanhuri, E. 2017. "Challenges of Construction of WtE Facility in Indonesia." Paper presented at the Workshop on Disaster Waste Management and Construction of WtE in Japan Environmental Sanitation Center (JESC), Bangkok, March 8.

Delgerbayar, Badam. 2016. "Solid Waste Management in Mongolia." Paper presented at the Seventh Regional 3R Forum in Asia and the Pacific, "Advancing 3R and Resource Efficiency for the 2030 Agenda for Sustainable Development," Adelaide, Australia, November 2–4. http://www.uncrd.or.jp/content/documents/4136Country%20 Presentation_Mongolia.pdf.

Dimishkovska, B., and J. Dimishkovski. 2012. "Waste Management in R. Macedonia." *QUAESTUS Multi Disciplinary Research Journal*. http:// www.quaestus.ro/wp-content/uploads/2012/03/dimiskovska42.pdf.

Dominican Republic, Ministry of Environment and Natural Resources. 2014. "Policy for the Integral Management of Municipal Solid Waste" ["Politica para la Gestion Integral de los Residuos Solidos Municipales"].

Ministry of Environment and Natural Resources [Ministerio de Medio Ambiente y Recursos Naturales]. Government of the Dominican Republic.

Dominican Republic, Ministry of Environment and Natural Resources and Ministry of Economy. 2017. Personal communication through interviews.

Ecuador, Ministry of Environment. 2018. National Program for Integrated Solid Waste Management [Programa Nacional para Gestión Integral de Residuos Sólidos], Ministry of Environment [Ministerio del Ambiente], Government of Ecuador. Personal communication with the World Bank, February.

Eisted, Rasmus, and Thomas H. Christensen. 2011. "Waste Management in Greenland: Current Situation and Challenges." *Waste Management and Research* 29 (10): 1064–70.

Enayetullah, I., A. H. Md. M. Sinha, and S. S. A. Khan. 2005. "Urban Solid Waste Management Scenario of Bangladesh: Problems and Prospects." Waste Concern.

Energy Answers. 2012. "Materials Separation Plan." Energy Answers Arecibo LLC., Energy Answers International Inc.

Eurostat. 2017. "Municipal Waste by Waste Operations [env_wasmun]." Accessed April 25, 2017. http://ec.europa.eu/eurostat/web/waste /transboundary-waste-shipments/key-waste-streams/municipal-waste.

Fiji, Department of Environment. 2011. "Fiji National Solid Waste Management Strategy, 2011–2014." Department of Environment, Ministry of Local Government, Urban Development, Housing and Environment, Government of Fiji. Accessed May 11, 2017. http://www .sprep.org/attachments/Fiji_NSWMS_2011-2014.pdf.

Frane, Anna, Asa Stenmarck, and Stefan Gislason. 2014. *Collection and Recycling of Plastic Waste: Improvements in Existing Collection and Recycling Systems in the Nordic Countries (Temanord)*. Copenhagen: Nordic Council of Ministers.

GIZ and SWEEP-Net. 2010a. "Country Report on the Solid Waste Management in Mauritania." German Corporation for International Cooperation [Deutsche Gesellschaft für Internationale Zusammenarbeit GmbH (GIZ)] and Regional Solid Waste Exchange of Information and Expertise Network in Mashreq and Maghreb Countries (SWEEP-Net), on behalf of the German Federal Ministry for Economic Cooperation and Development [Bundesministerium für wirtschaftliche Zusammenarbeit und Entwicklung (BMZ)].

———. 2010b. "Country Report on the Solid Waste Management in Syria." German Corporation for International Cooperation [Deutsche Gesellschaft für Internationale Zusammenarbeit GmbH (GIZ)] and Regional Solid Waste Exchange of Information and Expertise Network in Mashreq and Maghreb Countries (SWEEP-Net), on behalf of the German Federal Ministry for Economic Cooperation and Development [Bundesministerium für wirtschaftliche Zusammenarbeit und Entwicklung (BMZ)].

————. 2014a. "Country Report on the Solid Waste Management in Algeria." German Corporation for International Cooperation [Deutsche Gesellschaft für Internationale Zusammenarbeit GmbH (GIZ)] and Regional Solid Waste Exchange of Information and Expertise Network in Mashreq and Maghreb Countries (SWEEP-Net), on behalf of the German Federal Ministry for Economic Cooperation and Development [Bundesministerium für wirtschaftliche Zusammenarbeit und Entwicklung (BMZ)].

————. 2014b. "Country Report on the Solid Waste Management in Egypt." German Corporation for International Cooperation [Deutsche Gesellschaft für Internationale Zusammenarbeit GmbH (GIZ)] and Regional Solid Waste Exchange of Information and Expertise Network in Mashreq and Maghreb Countries (SWEEP-Net), on behalf of the German Federal Ministry for Economic Cooperation and Development [Bundesministerium für wirtschaftliche Zusammenarbeit und Entwicklung (BMZ)].

————. 2014c. "Country Report on the Solid Waste Management in Jordan." German Corporation for International Cooperation [Deutsche Gesellschaft für Internationale Zusammenarbeit GmbH (GIZ)] and Regional Solid Waste Exchange of Information and Expertise Network in Mashreq and Maghreb Countries (SWEEP-Net), on behalf of the German Federal Ministry for Economic Cooperation and Development [Bundesministerium für wirtschaftliche Zusammenarbeit und Entwicklung (BMZ)].

————. 2014d. "Country Report on the Solid Waste Management in Lebanon." German Corporation for International Cooperation [Deutsche Gesellschaft für Internationale Zusammenarbeit GmbH (GIZ)] and Regional Solid Waste Exchange of Information and Expertise Network in Mashreq and Maghreb Countries (SWEEP-Net), on behalf of the German Federal Ministry for Economic Cooperation and Development [Bundesministerium für wirtschaftliche Zusammenarbeit und Entwicklung (BMZ)].

————. 2014e. "Country Report on the Solid Waste Management in Morocco." German Corporation for International Cooperation [Deutsche Gesellschaft für Internationale Zusammenarbeit GmbH (GIZ)] and Regional Solid Waste Exchange of Information and Expertise Network in Mashreq and Maghreb Countries (SWEEP-Net), on behalf of the German Federal Ministry for Economic Cooperation and Development [Bundesministerium für wirtschaftliche Zusammenarbeit und Entwicklung (BMZ)].

———. 2014f. "Country Report on the Solid Waste Management in Occupied Palestinian Territories." German Corporation for International Cooperation [Deutsche Gesellschaft für Internationale Zusammenarbeit GmbH (GIZ)] and Regional Solid Waste Exchange of Information and Expertise Network in Mashreq and Maghreb Countries (SWEEP-Net), on behalf of the German Federal Ministry for Economic Cooperation and Development [Bundesministerium für wirtschaftliche Zusammenarbeit und Entwicklung (BMZ)].

———. 2014g. "Country Report on the Solid Waste Management in Tunisia." German Corporation for International Cooperation [Deutsche Gesellschaft für Internationale Zusammenarbeit GmbH (GIZ)] and Regional Solid Waste Exchange of Information and Expertise Network in Mashreq and Maghreb Countries (SWEEP-Net), on behalf of the German Federal Ministry for Economic Cooperation and Development [Bundesministerium für wirtschaftliche Zusammenarbeit und Entwicklung (BMZ)].

Global Methane Initiative. 2011. "Ethiopia Solid Waste and Landfill [Country Profile and Action Plan]." Community Development Research.

Gore-Francis, Janil. 2013. "Antigua and Barbuda SIDS 2014 Preparatory Progress Report." Environment Division, Ministry of Agriculture, Housing, Lands, and the Environment.

Greece, Ministry of Environment and Energy. 2015. "National Waste Management Plan" ["Εθνικο Σχεδιο Διαχειρισησ Αποβλητων"]. Ministry of Environment and Energy, Government of Greece.

Grenada, Population and Housing Census. 2011. http://finance.gd/images /Censussubmissionfinal.pdf.

Gret-LVIA-Pacte [Julien Rouyat, Cécile Broutin, Virginie Rachmuhl (Gret), Ahmed Gueye, Valentina Torrasani (LVIA), Ibrahima Ka (Pacte)]. 2006. "Household Waste Management in Secondary Cities of Senegal." Studies and Online Works # 8. ["La gestion des ordures ménagères dans les villes secondaires du Sénégal." Études et Travauxen ligne no. 8].

Guam. 2013. "Volume I: Guam Zero Waste Plan. Reaching for Zero: A Blueprint for Zero Waste in Guam." Government of Guam.

Guillaume, Marie, Bénédicte Château, and Alicia Tsitsikalis. 2015. "In-Depth Diagnosis of Waste Pre-Collection in Brazzaville" ["Diagnostic approfondi de la pré-collecte des déchets à Brazzaville"]. Groupe de Recherches et d'Echanges Technologiques.

Guyana, Ministry of Communities. n.d. "Putting Waste in Its Place: A National Integrated Solid Waste Management Strategy for the Cooperative Republic of Guyana, 2017–2030—Part 1: Our Strategy." Ministry of Communities, Government of Guyana.

Hong Kong, Environmental Protection Department, Statistics Unit. 2017. "Monitoring of Solid Waste in Hong Kong: Waste Statistics for 2016." Government of Hong Kong SAR, China.

Iceland, Statistics Iceland. 2015. *Statistical Yearbook of Iceland 2015* [*Landshagir 2015*]. Statistics of Iceland III, 108 [Hagskýrslur Íslands III, 108]. Statistics Iceland [Hagstofa Íslands], Government of Iceland, Reykjavík.

IDB (Inter-American Development Bank). 2012. "Strategic Plan for Solid Waste in Colombia" ["Plan Estratégico Sectorial de Residuos Sólidos de Colombia"]. IDB, Washington, DC.

———. 2013. "Water and Sanitation in Belize." Infrastructure and Environment Sector / Water and Sanitation Division (INE/WSA) Technical Note No. IDB-TN-609. IDB, Washington, DC.

———. 2015. "Status of Solid Waste Management in Latin America and the Caribbean" ["Situación de la gestión de residuos sólidos en América Latina y el Caribe"]. IDB, Washington, DC.

IDB-AIDIS-PAHO (Inter-American Development Bank, Inter-American Association of Sanitary and Environmental Engineering, and Pan American Health Organization). 2011. "Report of the Regional Evalution of the Management of Municipal Solid Waste in Latin American and the Caribbean 2010" ["Informe de la Evaluación Regional del Manejo de Residuos Sólidos Urbanos en América Latina y el Caribe 2010"]. IDB, AIDIS, and PAHO, New York.

IFC (International Finance Corporation). 2012. *Municipal Solid Waste Management: Opportunities for Russia*. Washington, DC: IFC Advisory Services in Eastern Europe and Central Asia.

IHSI. 2015. "Total Population, 18 Years and Above, Households and Densities Estimated in 2015." [Population Totale, de 18 Ans et Plus, Menages et Densites Estimes en 2015.] (Demographic and Social Statistics Branch, Haitian Institute of Statistics and Informatics [Direction des Statistiques Démographiques et Sociales, Institut Haitien de Statistique et d'Informatique]). March.

IHSI, IRD, Dial, Nopoor, ANR. 2014. "The evolution of living conditions in Haiti between 2007 and 2012. The social replica of the earthquake"["L'évolution des conditions de vie en Haïti entre 2007 et 2012. La réplique sociale du séisme"]. IHSI (Haitian Institute of Statistics and Informatics [Institut Haitien de Statistique et d'Informatique]), IRD (Research Institute for Development [Institut de recherche pour le développement]), and ANR (The National Research Agency [l'Agence nationale de la recherche]). Port-au-Prince, Paris, June.

IMF (International Monetary Fund). 2012. "Burkina Faso: Strategy for Accelerated Growth and Sustainable Development 2011–2015." IMF Country Report No. 12/123. IMF, Washington, DC.

India CPCB (Central Pollution Control Board). 2017. "Consolidated Annual Review Report on Implementation of Solid Wastes Management Rules, 2016 – Annual Review Report 2015–16." Government of India Ministry of Environment, Forests and Climate Change.

Intharathirat, R., and P. A. Salam. 2015. "Valorization of MSW-to-Energy in Thailand: Status, Challenges and Prospects." *Waste Biomass Valor* 7: 31–57. doi:10.1007/s12649-015-9422-z.

IPCC (Intergovernmental Panel on Climate Change). 2006. "Waste Generation, Composition and Management Data." In *2006 IPCC Guidelines for National Greenhouse Gas Inventories*, Volume 5: *Waste*. Edited by S. Eggleston, L. Buendia, K. Miwa, T. Ngara, and K. Tinabe. Kanagaw, Japan: Institute for Global Environmental Strategies.

Iraq, Ministry of Environment. 2015. "State of the Environment in Iraq 2015" [حالة البيئة في العراق لعام 2015]. Ministry of Environment, Government of Iraq.

Ireland, EPA (Environmental Protection Agency). 2014. "National Waste Report for 2012." EPA, Government of Ireland.

Isle of Man, Department of Infrastructure. n.d. "Waste Policy and Strategy, 2012 to 2022." Waste Management, Department of Infrastructure, Government of Isle of Man. https://www.gov.im/media/472034/waste_strategy.pdf.

Ismail, Anis. 2017. Personal communication between A. Ismail and Ministry of Environment.

Israel, Ministry of Environmental Protection. 2017. "Waste: Facts and Figures." http://www.sviva.gov.il/English/env_topics/Solid_Waste/landfilling/Pages/default.aspx.

Jamaica. 2011. "Number of Households by Method of Garbage Disposal by Parish." Census of Population and Housing, Kingston.

Jamaica NSWMA (National Solid Waste Management Authority). 2017. "Tonnage Islandwide 2016." NSWMA, Kingston.

Japan, Ministry of the Environment. 2015. "Waste Treatment in Japan FY2015 Version" ["Nihonno haikibutsu syori heisei 27 nendo ban"]. Ministry of the Environment, Government of Japan. http://www.env.go.jp/recycle/waste_tech/ippan/h27/data/disposal.pdf.

Kalula, Patrice Tshitala. 2016. "A Study of the Solid Waste Management Sector in the DRC: The Case of the City of Kinshasa, from 25 August to 19 December 2016" ["État des lieux de la gestion des déchets solides en République Démocratique du Congo: Cas de la ville de Kinshasa, situation du 25 août au 19 décembre 2016"].

KCCA-IFC (Kampala Capital City Authority and International Finance Corporation). 2017. "Project Teaser—Kampala Waste Treatment and Disposal PPP." Kampala Waste Management.

Keohanam, Bounthong. 2017. Director, Urban Development Division, Department of Housing and Urban Planning, Ministry of Public Works and Transport, Government of Lao PDR. Personal communication with the World Bank. May 8.

Korai, Muhammad Safar, Rasool Bux Mahar, and Muhammad Aslam Uqaili. 2017. "The Feasibility of Municipal Solid Waste for Energy Generation and Its Existing Management Practices in Pakistan." *Renewable and Sustainable Energy Reviews* 72: 338–53.

Kosovo, Ministry of Environment and Spatial Planning. 2013. "Strategy of the Republic of Kosovo on Waste Management 2013–2022."

Liechtenstein, Office of Statistics. 2018. "Liechtenstein in Figures 2018." Office of Statistics, Government of Liechtenstein.

LKSur. 2013. "Study of Characterization of Urban Solid Waste for Energy Purposes" ["Estudio de Caracterizacion de Residuos Solidos Urbanos con Fines Energeticos"]. ALUR (Alcoholes del Uruguay). MIEM DNE. February.

Macao SAR, China. 2014. "Report on the State of the Environment of Macao 2014." http://www.dspa.gov.mo/StateReportHTML/2014/pdf/en/04.pdf.

Macao SAR, China, DSEC (Statistics and Census Service). 2017. "Environmental Statistics 2016." DSEC, Government of Macao SAR, China. http://www.dsec.gov.mo/Statistic/Social/EnvironmentStatistics/20 16%E5%B9%B4%E7%92%B0%E5%A2%83%E7%B5%B1%E 8%A8%88.aspx?lang=en-US.

Macedonia, FYR, Ministry of Environment and Physical Planning. 2014. "State of Environment Report 2013." Macedonian Environmental Information Center, Skopje.

Mahapatra, Dhananjay. 2013. "Plastic Waste Time Bomb Ticking for India, SC Says." *Times of India*, April 4.

Maldives, MEE (Ministry of Environment and Energy). 2017. "State of the Environment 2016." MEE, Government of Maldives, Malé.

Mexico, SEMARNAT (Secretaría de Medio Ambiente y Recursos Naturales). 2016. "Report on the Situation of the Environment in Mexico" ["Informe de la Situación del Medio Ambiente en México"]. General Directorate of Environmental Statistics and Information, Secretariat of Environment and Natural Resources, SEMARNAT, Government of Mexico.

Modak, Prasad, Agamuthu Pariatamby, Jeffrey Seadon, Perinaz Bhada-Tata, Guilberto Borongan, Nang Sian Thawn, and Ma Bernadeth Lim. 2017. *Asia Waste Management Outlook*, edited by P. Modak. Nairobi: United Nations Environment Programme, Asian Institute of Technology, and International Solid Waste Association.

Moldova, Statistica Moldovei (National Bureau of Statistics of the Republic of Moldova). 2016. "Natural Resources and the Environment in the Republic of Moldova: Collected Statistics" ["Resursele naturale şi mediul în Republica Moldova: Culegere statistică" / Природные ресурсы и окружающая среда в Республике Молдова: Статистический сборник]. National Bureau of Statistics of the Republic of Moldova, Government of Moldova, Chişinău.

Monaco, Directorate of Environment. 2013. "The Environment in the Principality of Monaco" ["L'ENVIRONNEMENT en Principauté de Monaco"]. Directorate of Environment, Government of Monaco.

Naquin, Pascale. 2016. "Design and Implementation of a Quantification and Characterization Campaign for Household Waste in the Territory of AITOM le Marien (Cap Haïtien – Limonade – Quartier Morin)" ["Conception et réalisation d'une campagne de quantification et de caractérisation de déchets ménagers sur le territoire de l'AITOM le Marien (Cap Haïtien – Limonade – Quartier Morin)"]. Inter-American Development Bank, Washington, DC.

Nordic Competition Authorities. 2016. "Competition in the Waste Management Sector: Preparing for a Circular Economy." Nordic Competition Authorities, Konkurrensverket (Sweden), Konkurransetilsynet (Norway), Kilpailu ja Kuluttajavirasto (Finland), Samkeppniseftirlitid (Iceland), Konkurrence og Forbrugerstyrelsen (Denmark), Kappingareftirlitid (Faroe Islands), Forbruger og Konkurrencestyrelsen (Greenland). http://www.konkurrensverket.se/globalassets/publikationer /nordiska/nordic-report-2016_waste-management-sector.pdf.

OECD (Organisation for Economic Co-operation and Development). 2017. "Municipal Waste (Indicator)." OECD Data, OECD, Paris. https://data .oecd.org/waste/municipal-waste.htm.

Okot-Okumu, James. 2012. "Solid Waste Management in African Cities— East Africa." In *Waste Management: An Integrated Vision*, edited by Luis Fernando Marmolejo Rebellon, 3–20. London: InTechOpen.

Oriental Republic of Uruguay. 2004. "TOMO II: Urban Solid Waste. Solid Waste Management Plan of Montevideo and the Metropolitan Area" ["TOMO II: Residuos Sólidos Urbanos. Plan Director de Residuos Sólidos de Montevideo y Área Metropolitana"]. Office of Planning and Budgeting. Directorate of Development Projects.

Ouda, Omar. 2017. Civil and environmental consultant. Personal communication.

Papua New Guinea, NCDC (National Capital District Commission). 2016. "NCD Waste Management Plan 2016–2025: For a Sustainable Port Moresby." NCDC, Government of Papua New Guinea.

Patriamby, Agamuthu, and Masaru Tanaka. 2014. *Municipal Solid Waste Management in Asia and the Pacific Islands: Challenges and Strategic Solutions*. Singapore: Springer.

Peru, Ministry of Environment. 2013. "Report: "Solid Waste Diagnostics in Peru, Solid Waste NAMA Program" ["Informe: "Diagnostico de los residuos solidos en Peru, Programa NAMA de Residuos Solidos"]. Ministry of the Environment, Government of Peru.

———. 2016. "National Plan for Integrated Solid Waste Management, 2016–2024" ["Plan Nacional de Gestión Integral de Residuos Sólidos, 2016–2024"]. Ministry of Environment, Government of Peru.

PricewaterhouseCoopers Aruba. 2014. "Environmental Sustainability Ranking: Determining and Fortifying Aruba's Position in the Caribbean." Version 1.1. PricewaterhouseCoopers Aruba. https://www.pwc.com/an /en/publications/assets/pwc-environmental-sustainability-ranking-posi tioning-and-fortifying-aruba-in-caribbean.pdf.

Rebelde, Juventud. 2007. "The 100th Street Landfill" ["El Vertedero de la Calle 100"]. July 8.

Riquelme, Rodrigo, Paola Méndez, and Ianthe Smith. 2016. "Solid Waste Management in the Caribbean. Proceedings from the Caribbean Solid Waste Conference." Technical Note DB-TN-935. Inter-American Development Bank, Washington, DC.

Russia, Ministry of Natural Resources and Ecology. 2012. "The Substantiation of the Election of the Optimum Method of Waste Fund Disposal in the Cities of Russia." Federal Service for Supervision of Natural Resources Management Public Council, under Rosprirodnadzor the Commission of the Scientific Council of the Russian Academy of Sciences on Ecology and Emergency Situations. Accessed April 18, 2017.

San Marino, AASS (Autonomous State Company for Public Services). 2016. "Collection Data" ["Dati di raccolta"]. AASS, Government of San Marino. http://www.aass.sm/site/home/ambiente/dati-di-raccolta/2016 .html.

Saudi Arabia. n.d. "National Transformation Program 202." Program by Vision 2030, Saudi Arabia. http://vision2030.gov.sa/sites/default/files/NTP_En.pdf.

Shams, Shahriar, R. H. M. Juani, and Zhenren Guo. 2014. "Integrated and Sustainable Solid Waste Management for Brunei Darussalam." doi:10.1049/cp.2014.1066. https://www.researchgate.net/publication /272042749_Integrated_and_sustainable_solid_waste_management _for_Brunei_Darussalam.

SIDS DOCK (Small Island Developing States). 2015. "Toward the Development of a Caribbean Regional Organic Waste Management Sub-Sector: Development of a Caribbean Regional Organic Waste Management Conversion Sub-Sector to Increase Coastal Resilience and Climate Change Impacts and Protect Fresh Water Resources." Draft 1.0. SIDS DOCK Secretariat and Caribbean Community Climate Change Centre. May 18.

Singapore, Ministry of the Environment and Water Resources. 2017. "Key Environmental Statistics 2016." Government of Singapore.

Solomon Islands, MECDM (Ministry of Environment, Climate Change, Disaster Management and Meteorology). 2015. "Draft National Waste Management and Pollution Control Strategy 2017–2026: Honiara." Government of Solomon Islands.

South Africa, Department of Environmental Affairs. 2012. "National Waste Information Baseline Report: Draft." Department of Environmental Affairs, Government of South Africa, Pretoria, November 14.

SPREP (Secretariat of the Pacific Regional Environment Programme). 2016. "Cleaner Pacific 2025: Pacific Regional Waste and Pollution Management Strategy, 2016–2025." SREP, Apia, Samoa. https://sustainabledevelopment.un.org/content/documents/commitments/1326_7636_commitment_cleaner-pacific-strategy-2025.pdf.

Sri Lanka, Ministry of Mahaweli Development and Environment. 2016. "Comprehensive Integrated Solid Waste Management Plan for Target Provinces in Sri Lanka." Ministry of Mahaweli Development and Environment, Government of Sri Lanka.

St. Lucia Government Statistics Department. 2011. "2010 Population and Housing Census, Preliminary Report." Government of St. Lucia.

St. Vincent and the Grenadines, Statistical Office. 2012. "2012 Compendium of Environmental Statistics." Central Planning Division. Ministry of Economic Planning, Sustainable Development, Industry, Information and Labour. Government of St. Vincent and the Grenadines, Kingstown. http://www.cwsasvg.com/contactus.html.

———. n.d. "Population and Housing Census Report 2012." Statistical Office, Central Planning Division, Ministry of Finance, Planning and Economic Development, St. Vincent and the Grenadines, Kingstown.

States of Guernsey. 2017. "Guernsey Facts and Figures 2017." States of Guernsey Data and Analysis. https://www.gov.gg/CHttpHandler.ashx?id=110282andp=0.

SWANA Haiti Response Team. 2010. "Municipal Solid Waste Collection Needs in Port-au-Prince, Haiti: Position Paper." Solid Waste Association of North America (SWANA).

Takeda, Nobou, Wei Wang, and Masaki Takaoka, eds. 2014. *Solid Waste Management*. Urban Environment 3. Kyoto: Kyoto University Press.

Tanzania, NBS (National Bureau of Statistics) and OCGS (Office of Chief Government Statistician). 2014. "Basic Demographic and Socio-Economic Profile." NBS, Ministry of Finance, Dar es Salaam, and OCGS, Ministry of State, President's Office, State House and Good Governance, Zanzibar, Government of Tanzania.

Tas, Adriaan, and Antoine Belon. 2014. "A Comprehensive Review of the Municipal Solid Waste Sector in Mozambique: Background Documentation for the Formulation of Nationally Appropriate Mitigation Actions in the Waste Sector in Mozambique." Carbon Africa Limited, Nairobi, and Associação Moçambicana de Reciclagem, Maputo.

Thein, M. 2010. "GHG Emissions from Waste Sector of INC of Myanmar." Paper presented at the Eighth Workshop on GHG Inventories in Asia (WGIA8), Vientiane, Lao PDR, July 13–16.

Trinidad and Tobago, EMA (Environmental Management Authority). n.d. "State of the Environment Report, 2011." EMA, Government of Trinidad and Tobago.

Tsai, W. T., and Y. H. Chou. 2006. "An Overview of Renewable Energy Utilization from Municipal Solid Waste (MSW) Incineration in Taiwan." *Renewable and Sustainable Energy Reviews* 10: 491–502. doi:10.1016/j.rser.2004.09.006.

UFPE (Federal University of Pernambuco). 2014. "Analysis of the Various Treatment Technologies and Final Disposal of Urban Solid Waste in Brazil, Europe, the United States and Japan." ["Análise das Diversas Tecnologias de Tratamento e Disposição Final de Resíduos Sólidos Urbanos no Brasil, Europa, Estados Unidos e Japão"]. UFPE and National Bank for Economic and Social Development, Recife, Brazil.

UNCRD (United Nations Centre for Regional Development). 2017. "State of 3Rs in Asia and the Pacific Country Report: Malaysia," by Agamuthu Pariatamby. Secretariat of the Regional 3R Forum in Asia and the Pacific, UNCRD, Nagono, Japan, and Institute for Global Environmental Strategies, Kamiyamaguchi, Japan.

UNECA (United Nations Economic Commission for Africa). 2009. "Africa Review Report on Waste Management." Addis Ababa, October 27–30.

UNECE (United Nations Economic Commission for Europe). 2017. "Environmental Performance Reviews Series No. 46: Tajikistan." UNECE, New York.

UNFCCC (United Nations Framework Convention on Climate Change). 2014. Nkolfoulou Landfill Gas Recovery Project. Project Design Document. UNFCCC, New York.

UNSD (United Nations Statistics Division). 2013. "Dominica Environment Statistics Country Snapshot." UNSD, New York.

———. 2016. "Environmental Indicators: Waste. Municipal Waste Treated." UNSD, New York. http://unstats.un.org/unsd/ENVIRONMENT/qindicators.htm.

U.S. EPA (Environment Protection Agency). 2014. "Advancing Sustainable Materials Management: 2014 Fact Sheet: Assessing Trends in Material Generation, Recycling, Composting, Combustion with Energy Recovery and Landfilling in the United States." Office of Land and Emergency Management, EPA, Washington, DC.

———. 2016. "Sustainable Approaches for Materials Management in Remote, Economically Challenged Areas of the Pacific." EPA/600/R-16/303. EPA, Washington, DC.

van der Werf, Paul, and Michael Cant. 2007. "Composting in Canada." *Waste Management World*, January 3. https://waste-management-world.com/a/composting-in-canada.

Vietnam WENID (Waste Management and Environment Improvement Department). 2013. "Country Analysis Paper." Paper presented at the Fourth Regional 3R Forum in Asia, "3Rs in the Context of Rio+20 Outcomes: The Future We Want," Hanoi, March 18–20.

Viren, Lilta. 2013. "Transfer Station for Garbage at Saramaccakanaa" ["Overslagstation voor huisvuil aan het Saramaccakanaa"]. Anton de Kom Universiteit van Suriname.

World Bank. n.d. "Legal, Institutional, Financial Arrangement and Practices of Solid Waste Management Sector in Kazakhstan." World Bank, Washington, DC.

Zoï Environment Network. 2013. "Waste and Chemicals in Central Asia: A Visual Synthesis." Zoï Environment Network, with support from the Swiss Federal Office for the Environment. Accessed April 27, 2017. http://wedocs.unep.org/handle/20.500.11822/7538.

ZWMNE (Zero Waste Montenegro). 2016. "Waste Management Status in Montenegro." Accessed July 8, 2017. http://zerowastemontenegro.me /waste-management-status-montenegro.